D1554917

Springer Series in Language and Communication 18

Editor: W.J.M. Levelt

Springer Series in Language and Communication

Editor: W.J.M. Levelt

Volume 1 **Developing Grammars**
By W. Klein and N. Dittmar

Volume 2 **The Child's Conception of Language** 2nd Printing
Editors: A. Sinclair, R. J. Jarvella, and W. J. M. Levelt

Volume 3 **The Logic of Language Development in Early Childhood**
By M. Miller

Volume 4 **Inferring from Language**
By L. G. M. Noordman

Volume 5 **Retrieval from Semantic Memory**
By W. Noordman-Vonk

Volume 6 **Semantics from Different Points of View**
Editors: R. Bäuerle, U. Egli, A. von Stechow

Volume 7 **Lectures on Language Performance**
by Ch. E. Osgood

Volume 8 **Speech Act Classification**
By Th. Ballmer and W. Brennenstuhl

Volume 9 **The Development of Metalinguistic Abilities in Children**
By D. T. Hakes

Volume 10 **Modelling Language Behavior**
By R. Narasimhan

Volume 11 **Language in Primates: Perspectives and Implications**
Editors: J. de Luce and H. T. Wilder

Volume 12 **Concept Development and the Development of Word Meaning**
Editors: Th. B. Seiler and W. Wannenmacher

Volume 13 **The Sun is Feminine**
A Study on Language Acquisition in Bilingual Children
By T. Taeschner

Volume 14 **Prosody: Models and Measurements**
Editors: A. Cutler and R. D. Ladd

Volume 15 **Language Awareness in Children**
By D. L. Herriman

(continued after Index)

Donna-Lynn Forrest-Pressley
T. Gary Waller

Cognition, Metacognition, and Reading

Springer-Verlag New York Berlin Heidelberg Tokyo

Donna-Lynn Forrest-Pressley

Children's Psychiatric Research Institute
Box 2460
University of Western Ontario
London, Ontario, N6A 5C2
Canada

T. Gary Waller

Department of Psychology
University of Waterloo
Waterloo, Ontario, N2L 3G1
Canada

Series Editor

Professsor Dr. Willem J. Levelt
Max-Planck-Institut für Psycholinguistik
Berg en Dalseweg 79, Nijmegen, The Netherlands

Library of Congress Cataloging in Publication Data
Forrest-Pressley, Donna-Lynn.
Cognition, metacognition, and reading.
(Springer series in language and communication; 18)
Bibliography: p.
Includes indexes.
1. Reading, Psychology of. 2. Cognition in children.
3. Children—Language. 4. Memory in children.
5. Attention in children. I. Waller, T. Gary (Thomas
Gary), (date). II. Title. III. Series.

BF456.R2F67 1984 153.6 84-5402

With 1 Figure

Media conversion by World Composition Services, Inc., New York, New York.
Printed and bound by R.R. Donnelley & Sons, Harrisonburg, Virginia.
Printed in the United States of America.

9 8 7 6 5 4 3 2 1

ISBN 0-387-90983-4 Springer-Verlag New York Berlin Heidelberg Tokyo
ISBN 3-540-90983-4 Springer-Verlag Berlin Heidelberg New York Tokyo

Preface

We had our first conversation about cognition, metacognition, and reading in September of 1976. Our particular concern was with reading and learning to read, and what, if anything, *meta*cognition might have to do with it all. We didn't really know much about metacognition then, of course, but then most other people were in the same predicament. Some people had been working with interesting approaches and results on metalanguage and reading, among them J. Downing, L. Ehri, L. Gleitman, I. Mattingly, and E. Ryan, and it also was about that time that people were becoming aware of E. Markman's first studies of comprehension monitoring. Other than that perhaps the most influential item around was the perhaps already "classic" monograph by Kruetzer, Leonard, and Flavell on what children know about their own memory. Also in the air at that time were things like A. Brown's notions about "knowing, knowing about knowing, and knowing how to know," D. Meichenbaum's ideas about cognitive behavior modification, and the work by A. Brown and S. Smiley on the awareness of important units in text. Even though these developments were cited as new and innovative, it was not the case that psychologists had never before been interested in, or concerned with metacognitive sorts of questions. They certainly had, as clearly evidenced by the notion of "metaplans" in Miller, Galanter, and Pribram's *Plans and the Structure of Behavior*. But certainly, in our view at least, the general area of concern has been refreshed and re-energized in the last few years.

The ideas and the empirical investigation that we report here eventually grew out of these first conversations as we watched the area of "metacognition" emerge in its newer flavor. Several things seemed clear to us fairly early. First, any conceptualization of reading that includes only decoding and comprehension is inadequate. Why? Simply because skilled readers are able to read strategically—they can adjust what they do to the demands of the situation. Second, reading involves *meta*cognitive aspects as well as cognitive. That is, we don't just decode words; we also know about decoding. Skilled readers don't just comprehend; they monitor their comprehension and if something isn't working they do something about it. Skilled readers don't just read strategically; they know about and exert control over their strategic reading.

As we wondered about reading and its metacognitive aspects, and as we poked and prodded small groups of children with first one little question and then another, we quickly realized that we should not continue to look here and then there with small questions and small samples because each time we found something (with one set of kids) we could only wonder how it went with something else (in another set of kids). Rather, we should shoot the works, at least within the limits of good sense, available resources, and what little we knew about metacognition and reading. Our sense of things was that we could learn about how things go together and change over time if we went after a reasonably large set of children with a large, but hopefully systematic and rational, set of questions and measures.

What we eventually did is reported in detail here. (Parts of it have been reported earlier at conferences and workshops, but never the entire package.) Basically, we set out to examine, in some detail, the cognitive and metacognitive sophistication of 144 children with equal numbers from third and sixth grades and from children classified as poor, average, and good readers (i.e., 24 children in each grade by reading level combination). We attempted to find out what each child could do with respect to what we thought were the basic, important components of reading (i.e., decode, comprehend, and read strategically for a purpose) simply by giving them sets of *performance* tasks in each of the three areas. We also tried to find out what each child knew about him- or herself in each area by extensively interviewing each child individually (i.e., collecting their *verbalizations*) and, where we knew how, by giving them other clever things to do. Here we were particularly concerned, for each child, with whether he or she knew, for example, that there are alternative ways to do things (e.g., to decode), that you can monitor your own performance (as in comprehension monitoring), that if you aren't doing well perhaps you can try to reach your goal in some other way (i.e., select a new strategy or adopt a new plan), and so forth.

In a small fit of grandiosity we also decided to investigate not only reading, but also three other general areas of development (language, memory, and attention) which any prudent rational person would agree might be relevant to reading and learning to read. Specifically, we also examined what each of these 144 children could do on a broad array of performance tasks in these three areas, and we interviewed each of them to see if we could determine what they knew about their language, their memory, and their attention. While this grandiose decision doubled our task, our view is that the results are informative and provide a more complete view of the development of reading. We leave it to the reader, several years and many hours later, to assess for him- or herself what we learned from these very patient and cooperative children.

For any project of this sort it is inevitable that a large number of people and organizations have helped in countless ways. Funds to support this research were provided by the National Research Council (now the National Sciences and Engineering Research Council) and the Canada Council (now the Social Sciences & Humanities Research Council) of Canada, by the Research Development Fund at Dalhousie University and at Althouse College of the University of Western Ontario, by the Children's Psychiatric Research Institute, and by the Department of Psychology at the University of Waterloo. Special thanks go to R. Barron, G.E. MacKinnon, A. Cheyne, and E. Ryan, for their advice and critical comments at various times over the years. We are most grateful to D. Willows for her extensive and critical assistance, and for her help in arranging for us to work in the schools. Special thanks go to the principals, teachers, students, and parents of St. Francis and St. Gregory's Separate Schools, Cambridge, Ontario, and St. Francis Separate School, Kitchener, Ontario, for their cooperation in this study. We also wish to thank a number of other people: M. Tapley-Chiduck for helping with group testing and setting up the "blind" scoring procedure; Scholastic Publishing Co., Richmond Hill, Ontario, for helping with selection of reading materials; M. Laughlin for her patience and excellence in typing the interview transcripts; the Audiovisual Department, University of Waterloo, for solving technical problems with the audio tapes; M. Miller for doing the reliability checks; the students and colleagues who acted as raters for the stories and titles; M. Kohli and R. Crispin for the many hours of consultation over computer problems; G. Stevens for extensive work on some of the tables; N. Hultin for telling us how to spell "mimic" and its awesome derivations; and A. Bast simply for keeping things in order. We are particularly grateful to T. Cook for the many hours of work that it took to produce a final manuscript and for her expertise and good humor during it all. Others who helped with equal expertise and good cheer were C. Ledbury, P. Kovacs, and L. Reidel. There also are numerous colleagues who listened, suggested, and otherwise helped in so many ways. Finally, we must acknowledge those whose contributions are inevitably invisible but whose support and sustained encouragement are so vital. They are Michael (for DFP) and Jackie (for TGW).

Contents

CHAPTER 1

Introduction

Within the last several years, there has been a growing interest among psychologists and educators in the concept of *meta*cognition. According to Flavell (1976), metacognition refers to "one's knowledge concerning one's own cognitive processes and products or anything related to them." Metacognition also includes "the active monitoring and consequent regulation and orchestration of these processes in relation to the cognitive objects or data on which they bear, usually in the service of some concrete goal or objective" (Flavell, 1976, p. 232).

In other words, we have knowledge about our cognitive (i.e., mental) processes and we use this knowledge to choose the most efficient strategy for, or ways of dealing with, any problem that we might face. The particular problem could be as simple as remembering a telephone number or as complex as writing a research report. Regardless of what the task is, as we proceed we monitor and regulate our activities. For example, we monitor when we ask ourselves whether or not we have rehearsed a telephone number enough to be able to remember it. We might even use a self-test strategy (cover up the number and try to recite it) to check our progress. If we find that our goal has not been achieved, then we might again draw upon our knowledge of alternative strategies in order to rectify the situation. For example, we might realize that we have too many things to remember for that particular day and resort to writing down the telephone number to reduce the load. Alternatively, when writing a research report, we need to know about a number of very complex skills (e.g., reading text, identifying and extracting important information from a variety of sources, integrating and evaluating findings, and finally, actually writing, editing, and revising the required written document).

Flavell (1978) offered the following example of how metacognition interacts with cognition.

> For instance, we suddenly get the vague sensation (metacognitive experience)

> that we may not fully understand what we have just read, so we review (cognitive action) the material and our interpretation of it in order to find out exactly what, if anything, is amiss (another metacognitive experience). Or we may decide to read something for some purpose (establish a goal) and start by skimming parts of it (cognitive action) in order to get some initial sense of how hard the going is likely to be (metacognitive experience).

In short, metacognitive processes refer to the control or executive processes that direct our cognitive processes and lead to efficient use of cognitive strategies.

In spite of the fact that metacognition is a relatively "new" concept, it can be viewed within much the same framework as the model of cognitive processing proposed by many more traditional information-processing theorists. For example, Miller, Galanter, and Pribram (1960) argued that behavior is guided by the formation of "plans" (i.e., a hierarchy of instructions that controls the order in which a sequence of operations is performed). The mature individual has many more plans available than the one being executed and is capable of rapid alternation between plans. The individual also has images (i.e., accumulated, organized knowledge about self and world) that are used to select an appropriate plan. Miller and his colleagues suggest that learning only occurs when the person has some kind of a plan. Furthermore, a plan will not be achieved "without intent to learn, that is to say, without executing a *metaplan* for constructing a plan that will guide recall" (p. 129). It is those metaplans, then, that generate alternative plans. Once a plan is available, a control process referred to as a TOTE unit (test-operate-test-exit unit) guides behavior. This TOTE unit continually monitors the progress of the plan that is being activated. We believe that TOTE units and metaplans roughly correspond to the mechanisms of cognitive knowledge and control that mature readers use, and that plans correspond to specific strategies that can be activated by the higher-order cognitive control processes. It should be noted that the general Miller et al. model has influenced research in many domains such as language (e.g., Bloom & Lahey, 1978, p. 22), memory (e.g., Brown, 1978; Pressley, Heisel, McCormick & Nakamura, 1982), and intelligence theory (e.g., Sternberg, 1979).

Metacognition also might be an element common to all problem-solving tasks (Paris, 1978). According to Brown (1980), the ability to monitor one's cognitive processes is transsituational; it is a sign of efficient learning in many tasks. In addition, Brown and DeLoache (1978, p. 30) claimed that an "accumulation of knowledge about how to think in an increasing array of problem situations is an outcome of experience with more and more complex problems." If these assumptions are correct, then the acquisition of metacognitive skills may be not only a developmental issue, but also a matter of experience (cf., the novice–expert distinction in the problem-solving literature, e.g., Brown & DeLoache, 1978).

Other conceptions of the nature of metacognition also exist. For example, Bialystok and Ryan (in press) have described metacognition as the two disected planes of knowledge and monitoring. Both planes represent a continuum of complexity, and location of specific tasks in the matrix represents the interaction of the component planes and the rationale for developmental progress. On the other hand, Pressley, Borkowski, and O'Sullivan (in press) have suggested that metacognitive skills are the result of learning experiences and that metacognition results in procedures by which new cognitive strategies can be acquired.

Up to this point, we have discussed the concept of metacognition in purely theoretical terms. What, then, is the importance or relevance of this concept to the practitioner? It is quite possible that the metacognitive aspects of any task are an important component to include in training procedures, particularly if the assumptions of transsituationality are correct. Although we have a limited amount of information about metacognitive training procedures (e.g., Pressley, Borkowsi, & O'Sullivan, in press; Wong & Jones, 1982), we do have a large body of literature that advocates use of self-produced verbalizations during acquisition of skills (e.g., Meichenbaum & Asarnow, 1979). Also, metacognition might be the critical element that facilitates maintenance (continued use) and transfer (use of same strategy in different situations) of skills. There is limited evidence that use of a metacognitive component in instructional situations actually will improve the probability that a specific strategy will be generalized to a new situation (O'Sullivan & Pressley, 1983). Metacognition might be a major contributing factor to "learning to learn." (See Brown, Campione, and Day, 1981, and Pressley, Borkowski, and O'Sullivan, in press, for extensive discussions of this point.)

Unfortunately, one of the major difficulties in using the metacognitive framework in an applied situation is the confusion surrounding the theoretical constructs, and the lack of consistency in operational definitions. To date, there have been several attempts to clarify the concept of "metacognition" in various contexts, but there has not been a great deal of empirical work except for that by Flavell and his colleagues in the area of metamemory, and an emerging series of investigations of comprehensive monitoring (e.g., Baker, 1979; Baker & Anderson, 1982; Markman, 1977, 1979; Markman & Gorin, 1981) and identification of important units in text (e.g., Brown & Smiley, 1977). At best, it does not appear that the term metacognition has been used consistently by writers in the area, even at the conceptual level, not to mention the operational. At the operational level, it is not even clear that many of the tasks that have been used reflect metacognition as the term has been described here.

With respect to metacognitive research in general, there are at least two broad areas of concern. First, the relationship between cognition and metacognition is largely an unstudied area, even though it is a question of concern (Wong, in press). For example, Flavell and Wellman (1977) pointed out that there certainly should be a correlation between behavior in these two areas, but the actual relationship is hypothesized to be more interactive than linear in a causative sense, meaning that cognitions could cause metacognitions as well as vice versa. In addition, they argue that the various "metas" in metacognitive development may not emerge synchronously. The little empirical information that we do have about cognitive–metacognitive linkages is concerned chiefly with knowledge and strategy usage, usually in the area of memory (e.g., Borkowski, Reid, & Kurtz, in press; O'Sullivan & Pressley, 1983; Yussen & Berman, 1981). Moreover, the results are not always encouraging. For example, Cavanaugh and Borkowski (1980) assessed metamemory (through interview items developed by Kreutzer, Leonard, and Flavell, 1975) and memory performance in children in kindergarten, and grades one, three, and five. They reported significant correlations between items in which data were combined across grades, but within-grade correlations were not significant and did not generalize across memory tasks. That is, the amount of knowledge about strategies failed to distinguish individuals who used relevant knowledge from those individuals who did not. On the whole, the contention that successful metamemory is a necessary prerequisite for successful memory was not supported. However, those readers familiar with the world of test construction will realize that individual items typically are less reliable and have less predictive validity than a composite score based on any set of items. It is probable that a similar phenomena is at work in the data presented by Cavanaugh and Borkowski. If a composite score of memory performance was correlated with a composite score of memory knowledge, then the possibility of finding cognitive-metacognitive relations might improve. (See Rushton, Brainerd, and Pressley, 1983, for more information on the use of individual items in research on children's behaviors.) Certainly, the pattern of metacognitive development, both alone and in relation to cognitive development, needs to be more fully explored. It is hoped, too, that future research will try to make use of better statistical procedures whenever possible.

A second general concern of metacognitive research is that the role of metacognitive processes in reading and the relationship between cognitive and metacognitive aspects of reading have not been adequately examined. Although some have begun to explore the area (e.g., Brown & Smiley, 1977; Baker, 1979; Myers & Paris, 1978; Forrest & Barron, 1977), these studies, in most cases, have investigated reading skills either in terms of cognition *or* metacognition, but not both. If metacognitive skills tend to be task specific, as suggested by Flavell and Wellman (1977), then it would be reasonable and important to investigate a task, such as reading, that is important to the educational process, per se, and in which metacognition might play an active role. It is not unique to consider reading as an instance of complex problem solving, particularly when reading is considered in the context of an educational (learning) situation (e.g., Bransford, Stein, Shelton, & Owings, 1980; Kavale & Schreiner, 1979; Olshavsky, 1977; Reid, 1966). However, the use of reading as a problem- solving task with which to examine both cognition and metacognition does present several problems. (For an extensive discussion of how metacognition might interact with the reading process, see Forrest-Pressley & Gillies, 1983).

Perhaps one of the greater hindrances to the development of our understanding of reading derives from the fact that psychologists (and most others) have not been able to accept a common working definition (particularly an operational one) of the term "reading." For example, traditionally many psychologists and teachers have insisted that reading is nothing more than *decoding* written symbols to sounds (i.e., figuring out what the printed word says). Others traditionally have insisted that reading involves not only decoding from print to sound, but also *comprehending* the written material. The tendency on the part of many researchers, both historically and currently, has been to focus attention exclusively on one or the other of these two aspects of reading, or, at best, to accept the two as representing the totality of what we think of as "reading."

We feel that reading is not merely a decoding process, nor is it solely a comprehension process, nor is it just a "decoding plus comprehension" combination. Reading involves even more. It involves at least three types of skills: decoding, comprehension, and mature reading strategies (i.e., strategies of reading for a purpose). These types of skills or processes are all essential components of mature reading (e.g., Baker & Brown, in press; Brown, 1980; Gibson, 1972; Gibson & Levin, 1975; Forest-Pressley & Gillies, 1983; Rothkopf & Billington, 1975). Rather than deal with each type of skill separately, it seems important to consider reading as a complex system of skills and approach the problem with a series of converging measures.

When the mature reader decides to read, he usually has some purpose in mind, and the purpose for reading can vary considerably over situations (Gibson & Levin, 1975). For example, a person can read for entertainment or to find an answer to a problem. The reader might need to know only a specific piece of information, such as a telephone number, or in contrast, might need to understand a set of concepts presented by a writer. Whatever the purpose, the mature reader has available a number of strategies, such as rereading, skim reading, and paraphrasing, that can be used to help achieve a desired goal, whatever it might be. In effect, the skilled reader, in some sense, can read in different ways to meet different purposes. In this context, then, the first major concern of this investigation is reading *strategies,* a topic that has not been the subject of active research, particularly developmentally. The question we ask is, do mature readers adjust they way they read to the demands of a specific task? This concern originates from a body of literature exemplified by Rothkopf (1972) and Forrest and Barron (1977).

The second major concern of this investigation stems from the fact that reading traditionally has been considered, at base, a cognitive task, with a resultant primary concern for basic cognitive processes such as language, attention, and memory and their role in reading. An unfortunate consequence of such a view of reading is that there has been no room for concern for the "executive control" that is done by a skilled reader (e.g., the selection, monitoring, and modification of cognitive processes and strategies). One way to conceptualize and study these skills is in the framework of metacognition as described above. Only recently have several authors (e.g., Baker & Brown, in press; Brown, 1980; Brown &

Smiley, 1977; Forrest-Pressley & Gillies, 1983; Myers & Paris, 1978; Forrest & Barron, 1977) suggested that reading might involve metacognition as well as cognition.

As a result of the numerous approaches in the emerging field of metacognitive research, an obvious problem for this investigation was that of how to define, both theoretically and operationally, the terms metacognition and cognition. As the reader probably realizes by now, in the current state of the art, the only answers are rather undefined and imprecise. If a broad view of the study of cognitive and metacognitive development is taken, there appears to be some disagreement, or at least inconsistency, as to what constitutes "cognition," and in turn, what constitutes "metacognition." Given that there are no well-established guidelines in the field (particularly for metacognition), we rather arbitrarily adopted the following definitions. Cognition refers to the actual processes and strategies that are used by a reader. For example, when a child remembers something, memory processes per se are involved. When a child decodes a word, decoding processes per se are involved. On the other hand, *meta*cognition is a construct that refers, first, to what a person *knows* about his or her cognitions (in the sense of being consciously aware of the processes and being able to tell about them in some way), and second, to the ability to *control* (monitor) these cognitions (in the sense of choosing among alternative activities, and planning, monitoring, and changing activities). Metamemory, then, as the term is used here, refers to what a person knows about memory processes and is able to do about them (Flavell, 1977).

This notion of metacognition suggests a further extension to the concept of skilled reading. That is, skilled reading does not involve just decoding, comprehension, and reading strategies, per se. It also involves knowledge about each of these skills and the ability to control them (e.g., select, monitor, modify). For example, skilled readers can do more than simply decode a word. Skilled readers know that there are different ways to decode and can do something about their decoding activities (e.g., monitor them, change then, predict their adequacy). Again, mature readers can monitor their own reading comprehension and, if appropriate, modify reading activities to increase comprehension. Furthermore, the deliberate use of reading strategies should result in an increase in reading efficiency. In a sense, a mature reader is one who knows that he or she can read in different ways for different purposes and can do it appropriately. Specifically then, a second major concern of this investigation is the virtually unresearched area of *meta*cognition as it relates to reading and to the cognitive processes important to reading. That is, are there metacognitive aspects of reading?

The third major concern of this investigation is *development*. Are there differences, at different stages of development, in both the cognitive and metacognitive aspects of the different types of reading skills and the processes related to reading? Many attempts to create models of reading processes have been based to a large extent on data collected from adult subjects. Further, much empirical work on reading has been limited to a rather narrow range of ages. There has been insufficient concern for the development that occurs, particularly during the transition from beginning to skilled reading. It seems reasonable to assume that what a child brings, or does not bring, to a reading situation in terms of basic psychological processes is relevant to the development of reading skills. Although the catalogue of developmental factors that might be important to reading could be quite extensive, we chose for investigation the three basic areas of language, attention, and memory. No attempt is made here to justify the choice of these processes other than on obvious and intuitive grounds. Brief summaries of the relevant literature for each subset of skills is presented at the beginning of the appropriate chapters.

To summarize briefly, there are three major concerns in the investigation reported here: (1) the extent to which readers respond to the demands of different reading situations, (2) the role of metacognition in reading, and (3) the development of both cognitive and metacognitive aspects of reading and of basic processes important to reading. To address these three basic concerns, poor, average, and good readers in grades three and six were given an extensive set of tests to assess both their cognitive and metacognitive status regarding the components of reading (decoding, comprehension, and strategies) and processes related to reading (language, attention, and memory). We selected grades three and six because we felt that these children represented the difference between beginning and more mature reading. In other words, we felt that each of these grades represents a level in the development of the reading process, these being readers who are just beginnng to read fluently and readers who are fairly mature. Further, these grades were selected on the basis of results of previous research indicating that children in grades two and four were not using advanced reading strategies, whereas grade-six children were able to adjust their reading strategies to meet a specific purpose (Forrest & Barron, 1977).

Since the study adopted an expanded view of reading that included metacognitive as well as cognitive aspects of various processes and skills, it was essential to measure both aspects. First, it was necessary to assess the child's *use* of any particular skill. We refer to the tasks used to measure a child's use of a skill as *performance* measures, and these are the measures used to represent the cognitive aspect of any particular skill. For example, to measure cognitive aspects of memory, the children were given fairly typical, straightforward memory tasks. In many cases, the type of task used allowed the child to take advantage of a variety of strategies that, if used efficiently, could increase the level of performance. We felt that this flexibility, or appropriate use of skills, indicated the maturity of the skills involved. However, we did *not* make the assumption (occasionally made by other researchers) that flexible use of skills is an early indicator of metacognitive abilities. Rather, evidence of flexibility was taken to indicate only the maturity of cognitive skills. For example, with respect to decoding, the children were given materials to decode and scores on the specific tasks were used as an index of decoding performance, per se. We did not feel that it was important to know which decoding skill the child was using at any particular time; rather, we assumed that appropriate and flexible use of decoding skills would lead to increased performance.

In addition to measuring performance in various areas, it also was necessary to assess the child's knowledge of his or her cognitive processes (in the sense of being consciously aware of the processes and being able to talk about them). Here we were concerned with such things as the degree to which a person can make appropriate verbalizations about cognitive skills, about monitoring cognitive skills, and about using knowledge about these cognitive skills to predict efficiency in any given situation. This type of measure, for the most part, was taken in an interview session. (The one exception was a measure relating to reading comprehension, which results in scores reflecting a child's ability to predict the accuracy of his or her responses on a test. These measures were taken along with performance measures on tests of reading comprehension.) We refer to our attempts to assess what children know about their cognitive processes as *verbalization* measures because, for the most part, these measures were based on results of data collected in interviews. For example, children were interviewed about memory in much the same way that Kreutzer, Leonard, and Flavell (1975) studied metamemory. As a further example, children were interviewed about decoding, about the fact that we can decode in different ways and that different strategies might be more appropriate for different situations. From these interviews we developed measures of a child's metacognitive status as regards decoding.

Given the two types of measures described here (performance and verbalization), many researchers of metacognitive development quite willingly, without further concern, would equate these verbalization measures with measures of metacognition. However, we have some basic concerns that made us reluctant to do so. A major source of this reluctance is the realization that many of the skills about which we talked with the children were skills being taught in school. This is particularly true for the reading skills. Our guess is that it is possible for children to repeat, or "mimic," what a teacher has told them in class (perhaps repeatedly) without really understanding what is involved. If this is the case, then we feel that it is quite easy to seriously *overestimate* the metacognitive abilities of a child. (Also, see Nisbett and Wilson, 1977, for further discussion of the problems of relying on interview data.)

An alternative procedure is to be conservative and interpret verbal explanations in light of the actual use of any particular skill. Therefore, we decided to refer to items indicating the child's knowledge of cognitive processes (for the most part, the interview items) simply as verbalization items. Sophisticated metacognition was attributed to a child only when the child made appropriate verbalization about a skill or set of skills that the child *actually could use*. For example, a child might make rather sophisticated statements about decoding processes (i.e., about different ways to "figure out" what a word or sentence "said"), but not perform well at all on a decoding test. Our tact was not to attribute sophisticated metadecoding skills to that child. Our purpose in adopting this approach was to increase the likelihood that mature metacognition was attributed only to those children who really did understand their cognitive processes and could use this knowledge to monitor and modify performance.

Of course, we realize that this conservative definition, coupled with almost exclusive reliance on interview data, could lead to the *underestimation* of a child's metacognitive abilities. However, given the status of research in the area, our bias was to accept this risk. For those who do not prefer this conservative position, our verbalization data are reported inconsiderable detail and the reader can compare our conclusions with what a "traditional" researcher of "meta" might conclude. Even so, as the data are presented, the reader will realize that the more traditional view would lead to interpretations and conclusions that differ only slightly from ours.

To summarize again, we are concerned here with three things: (1) whether children respond to the demands of reading situations, (2) the role of metacognition in reading, and (3) cognitive and metacognitive development of reading skills and related processes. To address these concerns we tested a large sample of children on performance in various domains, and then interviewed them about their knowledge in these domains.

CHAPTER 2

General Method

The general overall plan of the investigation is summarized in Table 2-1. In simplest terms, the investigation was designed to assess the skills of poor, average, and good readers in grades three and six on both performance and verbalization items related to reading (decoding, comprehension, and strategies) and to developmental factors relevant to reading (language, attention, and memory). Thus, there are 12 "Sets" of data corresponding to the sections numbered one to twelve in Table 2-1.

The Children

The original population consisted of 227 children, 105 from third grade and 122 from sixth grade. All children were attending one of three suburban parochial schools in the Waterloo, Ontario, area. For every child, English was the first language. At the third-grade level, methods of teaching reading ranged from a strong phonics approach to a more eclectic approach that included phonics. (In general, spelling was being taught by a phonics method.) Comprehension skills were stressed to a lesser extent. At the sixth-grade level, the main emphasis was on comprehension skills, although decoding skills still were part of the reading program.

From the original group of 227, 144 children (72 from each grade) were selected to participate in the study. The children were selected at three reading levels by means of a group reading comprehension test, the Gates-MacGinitie Reading Test (Gates & MacGinitie, 1972), which has a mean standardized score of 50, with a standard deviation of 10. A poor reader was defined as a child who scored 45 or below (in standard score units), an average reader as one scoring between 45 and 55, and a good reader as one scoring above 55. Within each grade, 24 children were selected from each of these three reading levels.

Table 2-1. General Plan of the Investigation

	Grade and Reading Level					
	Grade 3			Grade 6		
	Poor	Average	Good	Poor	Average	Good
Reading Skills						
Decoding						
(1) Performance						
(2) Verbalization						
Comprehension						
(3) Performance						
(4) Verbalization						
Strategies						
(5) Performance						
(6) Verbalization						
Developmental Factors						
Language						
(7) Performance						
(8) Verbalization						
Attention						
(9) Performance						
(10) Verbalization						
Memory						
(11) Performance						
(12) Verbalization						

Thus, the children considered poor readers scored at about the 30th percentile or lower; those considered good readers were at about the 70th percentile or higher; the average readers were between the 30th and 70th percentiles. Sex distributions in the resulting six groups were as follows: the sixth-grade poor readers and third-grade average and poor readers each had 13 boys and 11 girls; the sixth-grade good readers included 11 boys and 13 girls; the other two groups included 12 of each sex.

The children also were given a nonverbal intelligence test, the Canadian Lorge-Thorndike Intelligence Test (Lorge, Thorndike, Hagen & Wright, 1967). Means and standard deviations for all of these premeasures (reading comprehension, chronological age, and nonverbal IQ) for each of the six major groups of children are presented in Table 2-2.

The following points are obvious from Table 2-2. (All statistical comparisons are based on analyses of variance followed by Newman-Keuls multiple comparisons.) (1) Standardized reading scores increased as a function of reading group, F (2, 138) = 381.86, $p < .001$, reflecting the fact that children were chosen to represent different reading levels. Overall reading scores for sixth-

Table 2-2. Means and Standard Deviations (SD) of Reading Comprehension, Chronological Age, and Nonverbal IQ

	Reading		Age		IQ	
	Mean	SD	Mean	SD	Mean	SD
Grade 3						
Poor	40.00	5.34	8.53	0.48	87.50	9.86
Average	50.25	2.54	8.29	0.56	95.79	13.86
Good	59.83	3.50	8.52	0.42	98.92	12.20
Grade 6						
Poor	41.75	2.56	11.47	0.36	86.50	11.65
Average	49.33	2.08	11.48	0.34	88.50	10.03
Good	62.96	4.67	11.47	0.37	104.58	14.10

grade children were slightly higher than for third-grade children, $F\ (1, 138) = 4.71$, $p < .05$. In addition, there was a significant interaction between grade and reading ability, $F\ (2, 138) = 3.81$, $p < .05$, such that there was no difference between poor readers in the two grades or between average readers in the two grades, but the sixth-grade good readers scored slightly higher than the third-grade good readers, $p < .01$. In effect, in terms of *relative* reading ability as reflected in the standardized reading scores, the sixth-grade good readers were slightly better than their third-grade counterparts. (2) There were no differences in age as a function of reading group for either grade, $F\ (2, 138) = 1.11$, $p > .10$, but obviously the sixth-grade children were older. (3) On the nonverbal IQ scores there was an interaction between grade and reading ability, $F\ (2, 138) = 3.46$, $p < .05$, in that in third grade the poor readers had significantly less ability than the average and good readers, $p < .05$, and in sixth grade the good readers had more ability than the poor and average readers, $p < .01$. (The nonverbal IQ results will not be reported again even though nonverbal IQ scores were included as a dependent variable in all subsequent multivariate analyses of variance of items.)

General Procedures for Data Collection

What follows is a general outline of the methodology used for the collection of the data. It is presented in detail for those interested in the design aspects of the experiment. For those less interested in the experimental details, it is sufficient to say that the data were collected in a standardized procedure, with the presentation of certain tasks being counterbalanced in order of presentation within sessions where appropriate. More detail is presented below.

Each of these 144 children included in the investigation received an extensive battery of test "items," including a standardized interview. Overall, the battery was designed to assess both the cognitive and metacognitive aspects of the basic components of reading (i.e., decoding, comprehension, and reading strategies) and of developmental factors related to reading (i.e., language, attention, and memory). For purposes of description and analysis, the items are categorized as performance and verbalization as shown in Table 2-1. The performance items included standardized as well as experimentally developed test items. The verbalization items consisted mainly of the interview questions (plus some "prediction-accuracy scores" calculated during the reading comprehension sessions). Some items were part of a group-administered test; other items, including the interview items, were administered individually. Some of the items actually were scores derived from a combination of items (e.g., the vocabulary score was based on a collection of words, a total decoding score was based on decoding of a set of words, etc.).

To avoid redundancy, most of the procedural details relevant to the administration of each item are presented in the various results sections along with the resulting data for the individual items. Only a general overview of the procedure is presented here. In brief, the overall plan of the test sequence was the same for all children and was as follows. First, the performance tasks assessing the cognitive aspects of language, attention, and memory (Sets 7, 9, and 11, Table 2-1) were presented in the form of group tests administered in the classroom to groups averaging in size from approximately 10 to 40 children. These group sessions lasted about one hour. The first individual session was an interview session in which most of the verbalization measures were taken (Sets 2, 4, 6, 8, 10, and 12, Table 2-1). The questions about the component processes of language, attention, and memory (Sets 8, 10, and 12) always were asked before the questions about decoding, comprehension, and strategies (Sets 2, 4, and 6). All possible random orders of questioning about language, attention, and memory were used equally often within each combination of grade and reading ability, but the sets of questions about decoding, comprehension, and strategies always were asked in the same order (decoding, strategies, comprehension). The reading questions were asked in the order of decoding, strategies, and comprehension because it was felt that the questions involved a natural progression. Within each of the sets (language, attention, memory, decoding, comprehension, and strategies), all the questions were given in the same order because it was felt, once again, that the questions involved a natural progression. In each of the results sections, all interview items are presented in the order in which they were asked of the children. The decoding performance tasks (Set 1, Table 2-1) all were done at the end of the first individual session, which usually lasted about 1 hr. The two remaining sessions were the reading sessions, and were concerned with comprehension and advanced strategies (Sets 3 and 5, Table 2-1). (The prediction-accuracy measures relevant to Set 4, Table 2-1, also were taken during these sessions.) These reading sessions lasted from 30 min to 1 hr.

All individual sessions were conducted in a quiet room in the child's school. In each case, the child sat opposite the tester, with a small table between them.

During the interview, a tape recorder was placed on the table and a small microphrone was hung around the child's neck. Before each session, the tester briefly established rapport with the child and then explained the purpose of the particular session. It was emphasized that there were no "right" answers for questions asked during the interview, and sometimes there was no "real" answer.

Scoring and Analyses

Performance Items

The scoring procedures for the various performance tasks are given in the appropriate section (with details presented in various appendices, which are referred to in the appropriate section). The performance items (with the exception of the measures for reading comprehension and strategies) were analyzed by using multivariate analyses of variance (MANOVA), in which reading level and grade were the independent variables and all the dependent variables in a set (e.g., all the decoding items) were analyzed simultaneously. Subsequent univariate analyses of variance on individual items were done where appropriate (i.e., when a significant MANOVA effect was observed). All subsequent comparisons of means were based on Duncan analyses, at a probability level of .05. It seemed reasonable to use a relatively liberal test for comparisons (such as the Duncan) since very little interpretive weight was being placed on the results of any one item. In addition, since nonverbal IQ is correlated with most of the variables being used in this investigation, it is possible that a group of items (excluding nonverbal IQ) could produce a significant MANOVA effect as a result of this confounding variable. Therefore, nonverbal IQ was included as a dependent variable in all MANOVA analyses. This obviously did not control for the effect of nonverbal IQ, but it did attenuate the weights attributed to the other items in the MANOVA. However, it also is the case that including a variable, such as nonverbal IQ, known to correlate with reading ability could have increased the likelihood of obtaining an overall significant F value. If nonverbal IQ produced the only significant univariate F values, this might have been a concern. However, considering the results of the analyses that included nonverbal IQ as a dependent variable, a decision was made to include rather than exclude nonverbal IQ as a dependent measure. In addition, to assess the relationship between each item and reading ability at each grade, correlations and partial correlations (controlling for nonverbal IQ) were computed. In all cases, there were 70 degrees of freedom for the correlations and 69 degrees of freedom for the partial correlations.

In the case of reading comprehension and reading strategies, the performance measures constituted a separate factorial design embedded in the overall plan (this is fully outlined in the appropriate section). These performance measures were analyzed by means of analyses of variance and covariance, with nonverbal IQ as the covariate. All subsequent comparisons of means were based on analyses by the Newman-Keuls procedure, at a probability level of .05. This method of comparison was used whenever it was more important for reasons of interpretation to be relatively confident about observed differences.

Verbalization Items

Responses to the interview (verbalization) items were scored on a simple 0-1-2 basis; 0 meaning a wrong response or no idea about what to do, 1 meaning some appropriate information was given, and 2 meaning information was given in a complete and sophisticated manner. Complete scoring criteria are given with each item in appendices referred to in the various results sections. All responses were recorded on tape, transcribed verbatim to typed script, and then scored using a "blind" procedure (i.e., the scorer did not know the grade or reading level of any child when the interview response was being scored). One half of the interviews (randomly selected with the constraints of equal numbers from each grade, reading level, and order of presentation of sets of questions) were scored independently by a second rater. Interrater reliability, then, was calculated as the percentage of items in the sample of response protocols for which the two scorers gave identical scores. (These figures are reported for all verbalization items in Appendix A.) In general, agreement between the two scorers was quite high on all items. Agreement was above 90% on 21 items, between 80 and 89% on 24 items, between 70 and 79% on 4 items, and at 68% on the one remaining item. When there was disagreement, the difference usually was only one point. On the 50 items scored for 72 children, differences of two points in the scoring scheme occurred only 0.64% of the time. In the case of disagreement, the judgment of the first scorer was accepted. Chi-square analyses were done on the resulting frequencies of children receiving each score (0, 1, 2) in each grade and at each reading level. For all of these initial chi-square analyses, there were ten degrees of freedom. If the overall chi-square value indicated that there was a significant relationship involving grade, reading ability, and the level of response, the item was analyzed further in an attempt to identify the specific "location" of the effect. To do this, the frequency of "sophisticated" responses (scores of 2) were compared with the frequency of "less mature" responses (scores of 0 and 1 collapsed or totalled). Subsequent analyses were based on chi squares done for each of the following groups: (1) third grade versus sixth grade (combined over reading groups), (2) all three reading groups (combined over grade), (3) poor versus average readers (combined over grade), (4) average versus good readers (combined over grade), (5) each two adjacent reading groups at each grade, and (6) third-grade good versus sixth-grade poor readers. All statistically significant differences that were found are reported with the appropriate item. All chi squares from these two-by-two contingency tables (i.e., where there is one degree of freedom) were corrected for attenuation using the Yates procedure. In addition, to assess the relationship between each verbalization item and reading ability, correlations and partial correlations (controlling for nonverbal IQ) were computed. In all cases, there were 70 degrees of freedom for the correlations and 69 degrees of freedom for the partial correlations.

Computed Scores

In order to examine the *overall* relationship between performance and verbalization, it was desirable to compute a single 'global' score that would indicate an individual's overall level of achievement on performance items for each factor and another score that would indicate an individual's overall level of achievement on verbalization items for each factor. For example, it was considered highly useful to compute a single score for each child on decoding performance (Set 1, Table 2-1) and on decoding verbalization (Set 2, Table 2-1). In order to compute the desired scores, all relevant item scores were normalized (e.g., converted to standard scores). That is, on each item the mean raw score over 144 children was computed and the individual raw scores were converted to z scores (to ensure equal weighting of each item). Then for each child a mean z score was computed for all items in a set (e.g., in decoding performance). Then a constant of 10 was added to each mean z score to eliminate negative values. The end result was a set of 144 mean z scores (plus 10), one for each child, on each of the 12 sets indicated in Table 2-1. Throughout the book, these mean z scores are referred to as *computed scores*. In cases where there were missing data (i.e., when, due to tester error, a child had not been asked a particular question), the mean z score for that grade and reading level was substituted before arriving at the computed score. Missing data accounted for less than 0.30% of the entire data matrix. Each computed score (e.g., decoding performance) was analyzed in a one-way analysis of variance. In addition, regression equations to predict reading ability at each grade were computed using the following combinations of scores: (1) performance and verbalization scores for each factor (e.g., language performance and language verbalization), (2) all computed scores representing reading skills (Sets 1 through 6, Table 2-1), (3) all computed scores representing developmental processes (Sets 7 through 12, Table 2-1), (4) all performance computed scores (all odd-numbered Sets, Table 2-1), (5) all verbalization computed scores (all even-numbered Sets, Table 2-1), and (6) all computed scores. In each case, nonverbal IQ was included in the regression equation. This series of regression equations was computed to examine the relationships between subsets of the data that may be seen as important to researchers in various fields. In addition, the reliability of each regression equation was assessed by splitting the data in half and redoing each regression equation for each half of the data. The results (order of entry into the equation for each variable for each equation) for the complete data set, half (A), and half (B) are presented in Appendix B. As can be seen there, the regressions are more reliable at the higher grade and when fewer variables are involved. However, the general pattern of results is confirmed. (For the regression equations, the F value and the tolerance level were not restricted, $F = .01$, $T = .001$, so that the most complete output could be obtained. However, this also meant that many variables were entering the equation when they in fact accounted for very little of the additional variance. All F values for the complete analyses are also presented in Appendix B.) Correlations and partial correlations, controlling for nonverbal IQ, also were computed to assess the relationship between each computed score and reading ability at each grade.

"Meta" Categorizations

As indicated earlier, there was concern that mature metacognition should be attributed to a child only when both performance and verbalization scores were high. As a reflection of this concern, children were categorized on the basis of the computed scores in the following way. Since 10 had been added to a set of standard z scores in the original computation, a score of 10 was taken as the midpoint, or critical cutoff score, in each set. A child who scored high (above 10) on both performance and verbalization tasks on any two related sets (e.g., decoding performance and decoding verbalization) was classified as META. In effect, this child was considered to be performing above average on both performance and verbalization items. The inference was that the child could use the cognitive skills *and* knew about the cognitive processes involved. A child who scored high on performance and low on verbalization (i.e., lower than 10) was considered as in TRANSITION. The inference was that the child could do the cognitive tasks (at least better than average), but lacked the sophistication necessary to talk about the processes. A child who scored low on performance and high on verbalization was classified as a MIMIC. The inference was that the child could mimic appropriate verbal responses when asked about cognitive processes (better than average), but could not do the performance tasks and did not use the information to increase performance. Finally, a child who scored low on both performance and verbalization was considered LOW. The inference here was that the child could neither use nor talk about the cognitive skills.

The resulting categorization frequencies were analyzed by means of chi squares. If the overall chi-square value indicated a statistically significant relationship involving grade, reading ability, and the categorization frequencies, then the categorization was analyzed further. To do this, the number of children being classified as META were compared with the combined numbers of children classified in "less mature" categories (LOW, MIMIC, and TRANSITION). These subsequent analyses were based on chi squares done for the following groups: (1) grade three versus grade six (combined over reading ability), (2) all three reading groups (combined over grade), (3) poor versus average readers (combined over grade), (4) average versus good readers (combined over grade), and (5) each two adjacent reading groups at each grade, plus third-grade good versus sixth-grade poor readers. All statistically significant results are reported in the appropriate sections. Again, all chi squares from these two-by-two contingency tables (i.e., where there was one degree of freedom) were Yates corrected.

In summary, each set of data was analyzed separately and various sets were combined in several ways. The structure of these analyses will be clear as the various sets are discussed. In the following chapter we examine decoding in detail.

CHAPTER 3

Decoding

Decoding: Cognition

Teachers and parents of young children usually know that when children learn to read, one of the things they have to learn is to "sound out" the words, or "figure out what the words *say*." The relative importance of this sounding out, or "decoding" as it is called, and the ways that schools should handle it has been a topic of considerable controversy. It even has been the focus of one of the "great debates" in twentieth-century education (Chall, 1967), with the "phonics" champions on one side and the "look say," or "whole word," advocates on the other. While the phonics group insists that decoding is a matter of pairing sound with symbol, the whole-word advocates claim that, since in our language there is not a simple or one-to-one correspondence between letters and sounds, the most efficient means of decoding is by recognizing words as "whole" words or as whole visual patterns.

Unfortunately, even this simple distinction is an over-simplification of the numerous and complex issues involved in the general area of decoding skills. Indeed, it would be presumptuous (unnecessary) for us to attempt to summarize the major works concerned with decoding. Therefore, the reader is referred to the many excellent reviews of the literature (e.g., Barron, 1981; Gibson & Levin, 1975; Golinkoff, 1976).

In any case, most children do learn to decode written symbols. In fact, and of considerable importance, it appears that children even are able to adjust their decoding activities and strategies to correspond to the ways in which they are taught (Barr, 1975). That is, not only do most children learn to decode words, they also appear to learn to do it in a variety of ways depending, at least to some extent, on what their teachers ask of them. Thus, it appears reasonable to assume that children generally do not have any particular cognitive limitation in learning decoding skills, and in fact are able to adjust their decoding strategy, at least to some extent, to meet the needs of the situation. This ability to adjust to the situation is of considerable interest to us.

Decoding: Metacognition

As suggested earlier, we want to argue that decoding can be accomplished in several different ways. For example, a child might recognize a word as a whole because he has seen it many times. Or, alternatively, the child might sound it out, "guess" its identity from the context, or ask someone what the word is. If a child has all of these possibilities within his or her repertoire, then at any given time any one of them might be used in a decoding task. The child might even *choose* which approach to use. This choice, in our view, reflects strategic use of the available decoding skills. In addition to making a choice, the child might actually be *aware* of the limitations of each decoding method. If so, then when faced with a specific set of circumstances, he or she might be able to optimize performance by choosing the most efficient strategy or approach for the specific situation. For example, if a child does not recognize a word as a whole and realizes that "figuring out what a word says" might help comprehension, he might try to sound it out. If he is unable to sound it out and realizes that he usually guesses poorly, and that there is no one else to ask, then he might ask the teacher what the print says.

To summarize to this point, we are using the word *decoding* here to refer, in a sense, to the acts or processes of translating written symbols into sounds. These translation processes can be supported by a number of strategies such as direct word recognition, sounding out, asking someone, using a dictionary, and guessing. *Meta*cognitive aspects of decoding, as we are using the term, refers to the idea that a child knows that there are different ways in which to figure out what a word says, that for any particular situation some decoding strategies are more efficient and appropriate than others, and that the child can evaluate the situation, assess the likelihood of dealing successfully with the situation in different ways, choose ways to approach the task, assess the adequacy of performance, and modify behavior if appropriate.

Given these rather abstract definitions, the next question is how to operationalize them, or how actually to measure them. Cognitive aspects of decoding can be assessed simply by determining how well a child performs on decoding tasks. Metacognitive aspects of decoding can be measured by assessing a child's use of these decoding strategies, plus assessing his or her ability to verbalize the different strategies and predict their efficiency. With these basic ideas in mind, the children in our study were asked to do several decoding tasks and were interviewed about their knowledge of decoding.

Table 3-1. Decoding Performance Items: Means

Item	Maximum Score	Grade 3			Grade 6		
		Poor	Average	Good	Poor	Average	Good
D-P-1	100	26.70	39.50	52.50	57.90	71.80	87.30
D-P-2	30	9.25	15.25	20.25	18.13	22.88	28.00

Method, Results, and Discussion

Performance Items

The cognitive aspects of decoding were assessed by a set of performance items (Set 1, Table 2-1). Two types of decoding performance items were used (and are presented fully in Appendix C). The first type used was the Slosson Oral Reading Test (Slosson, 1963). This test was used because it contains lists of words graded for decoding difficulty (e.g., list 1: came; list 4: serious; list 8: intangible) that could be decoded by direct word recognition, sounding out, or guessing. It was scored out of a possible maximum of 100 points. It is referred to as item D-P-1 in subsequent tables (i.e., D for decoding, P for performance, and 1 for first item in the Set).

The second set of items was based on the oral reading of nonsense syllables (e.g., clup, sart), taken from the Alta-Boyd Phonics Test (unpublished) and referred to in the tables as D-P-2. The maximum possible score was 30. This test assesses the extent to which phonics generalizations have been internalized to the point that they can be applied in *unfamiliar* contexts. The addition of the list of nonsense syllables insured that the children would be more likely to use a sounding-out strategy to be efficient in this decoding situation. Another possible, but less efficient, strategy that could have been used was guessing. That is, since the items were nonsense syllables, it was considered very unlikely that the children ever would have seen the items. Thus, it would not have been possible for them to decode them by direct word recognition.

Average scores (means) for each of the two items are presented in Table 3-1. All summaries of statistical analyses of the decoding items are shown in Appendix D. Overall, as is obvious from Table 3-1, performance on decoding items increased with grade and with reading ability. In addition, as shown by univariate analyses of variance, performance on *each* decoding item increased with grade and with reading ability. Individual comparisons showed that poor readers did less well on the decoding tasks than average readers, and average readers did less well than good readers.

The relationship between each performance item and reading ability at each grade level was assessed by means of correlations and partial correlations (controlling for nonverbal IQ). The results (shown in Table 3-2) indicated that the

Table 3-2. Decoding Performance Items: Correlations With Reading Scores

	Grade 3		Grade 6	
Item	*r*	partial *r*	*r*	partial *r*
D-P-1	.8012*	.7675*	.7874*	.7331*
D-P-2	.6435*	.5935*	.6464*	.6060*

$^*p < .001$

relationship between decoding performance and reading ability was quite strong (in the 0.6–0.8 range), even when the variance attributable to a nonverbal IQ was controlled.

The analyses of the decoding performance data suggest that younger/poorer readers generally have some difficulty with the basic reading skills of decoding. It is important here to note the low *absolute* levels of performance on these decoding tasks. Since these children had been exposed to methods of teaching that encouraged a "sounding out" strategy, it was assumed that all of them would have a reasonable chance for success on the nonsense words, if not on the Slosson reading test. However, as can be seen in Table 3-1, the grade three poor and average readers performed at relatively low levels. It also is interesting to note that there was no difference between third-grade good and sixth-grade poor readers on either decoding task. That is, in spite of differences in age and in length of time in school, these two groups appeared to be using decoding skills at about the same level of proficiency.

These two decoding performance items were used to compute a single overall decoding score for each child that would indicate the child's overall level of achievement on decoding performance as described in the General Method section. These computed scores on decoding performance, presented in Table 7-1 in a later chapter, increased with grade and with reading ability but did not interact significantly (see Appendix D). More will be said about these scores later.

Verbalization Items

In this section, the procedure and results for items concerned with the ability to make appropriate verbalizations about one's own decoding processes are reported. These items (Set 2, Table 2-1) are referred to as verbalization items and the data were collected in the interviews. The complete items and scoring criteria are presented in Appendix E. Each item is coded (e.g., D-V-1 refers to decoding, verbalization, item 1). Some verbalization items are similar to questions used by Myers and Paris (1978) but were developed independently. In general, items that were similar in the two studies produced similar results. However, since this investigation used a simple 0-1-2 scoring system rather than a category

Table 3-3. Decoding Verbalization: Frequency of Scores

		D-V-1	D-V-2	D-V-3	D-V-4
Grade 3					
Poor	0	2	8	2	5
	1	12	8	19	15
	2	10	8	3	4
Average	0	0	10	5	4
	1	10	3	15	15
	2	14	11	4	5
Good	0	0	3	2	3
	1	11	7	19	16
	2	13	14	3	5
Grade 6					
Poor	0	0	4	1	0
	1	3	4	18	16
	2	21	16	5	8
Average	0	0	2	2	1
	1	5	2	16	16
	2	19	20	6	7
Good	0	0	1	2	0
	1	5	0	20	9
	2	19	23	2	15
Chi square		25.05*	35.20**	7.88	24.22*

*p < .01
**p < .0001

system (as used in the Myers and Paris study), it is difficult to compare the results more directly.

The children were asked the following questions in order to assess their knowledge of strategies for decoding single words and sentences, and to assess whether or not they were aware that decoding by sound might not be sufficient for comprehension. In addition, there was an attempt to have the child express the point that some decoding strategies might be more efficient than others. The number of children in each grade and reading level giving any particular level of response for each verbalization item is shown in Table 3-3. For example, among the 24 sixth-grade good readers on item D-V-1, 19 received a score of 2, 5 received a score of 1, and none received a score of 0. (The appropriate overall chi-square values and probability levels are shown in the table in Appendix D.)

Question D-V-1: What do you do when you are reading and you come to a word that you don't know?

Most children, even the younger/poorer readers, could think of and talk about at least one way to figure out what a word "says." Chi-square analyses showed that the number of strategies that were mentioned by any particular child tended to increase with grade (but not with reading level). In addition, third-grade good readers reported fewer strategies than sixth-grade poor readers.

Question D-V-2: Is there a difference between knowing what a word says and knowing what a word means?

This item tended to differentiate between the various groups to a considerable degree. The ability to recognize the difference between "says" and "means," and then to give an adequate explanation of the difference, increased significantly both with grade and with reading ability.

Question D-V-3: Is it better to sound out a word that you don't know or to ask someone what it says? [Pause] Why?

When asked this most children suggested that sounding out was the better strategy, but went no further in their response. It is quite possible that this preference is due to the teaching method being experienced by these children rather than to any internal reasoning done by the child. This item did not discriminate among reading or grade levels.

Question D-V-4: What would you do if you were reading and you came to a whole sentence that you couldn't understand?

The ability to produce more than one strategy for figuring out what a sentence says increased mainly with grade. In addition, the difference in the frequency of immature versus sophisticated responses between sixh-grade average and sixth-grade good readers was significant.

The relationship between scores on each of the decoding verbalization items and reading ability at each level also was assessed by means of correlations and partial correlations controlling for the variance attributable to nonverbal IQ (see Appendix D). The only item that maintained a strong relationship to reading ability, with nonverbal IQ controlled, was the item that asked the children to differentiate between "what a word says" and "what a word means" (Question D-V-2). The partial r was .29 for third-graders and .25 for sixth-graders. More will be said about this later.

In addition to the quantitative differences substantiated by statistical analysis, there are qualitative differences among children at the various levels that can only be seen by actually reading the verbatim transcripts. For this reason, excerpts from selected good and poor readers in each grade will be presented throughout the book. We hope that these excerpts from the transcripts will give a clearer sense of what the child, at any particular level, is actually thinking.

The results of the statistical analyses of the decoding verbalization items suggest that, on the whole, the major difference among children would be a grade difference rather than a difference due to reading ability. Moreover, throughout the decoding interviews it is possible to see the effects of teaching and experience that would contribte to the grade effect. This is particularly true for poor readers. For example, one *third-grade poor reader* responded as follows to the interview questions. (In what follows, Q represents a summarized version of the question, while A is the actual verbatim reponse of the child.)

Q. What do you do when you come to a word that you don't know?
A. I sound it out.
Q. Anything else?
A. That's it.
Q. Is there a difference between knowing what a word 'says' and what a word 'means'? When you knew what it said, would you know what it meant?
A. Sometimes.
Q. Sometimes and sometimes not? How come sometimes not?
A. Sometimes I read the word wrong and then it doesn't mean right.
Q. But if you had read the word right, then you would know what it meant?
A. Yeah.
Q. You would? OK.
Q. Is it better to sound out a word you don't know or ask somebody what it says?
A. Sound it out and if you don't know the word, after you sound it out, you should ask.
Q. Why that way?
A. At least you tried to sound out the word or try to get the words.
Q. Why is it better to sound it out, do you suppose?
A. 'Cause you don't have to bother any kids.
Q. What do you do when you come to a whole sentence that you do not understand?
A. No idea.

This third-grade poor reader realized that you can sound out words or you can ask someone to decode them for you. However, she argued that it was better to try to sound out words than to "bother any kids." If a word was decoded properly, then the meaning would be obvious. Decoding sentences rather than words, however, was a problem that could not be fathomed. Also at the same grade level, slight differences can be seen between reading abilities. For example, one *third-grade good reader* said the following:

Q. Reading and come to a word that you don't know.
A. I sound it out.
Q. Anything else you can do?
A. I ask a friend what it is.
Q. Anything else?
A. I guess that's it.
Q. Difference between knowing what a word says and what it means. If you knew what it said, would you know what it meant?

A. Sometimes.
Q. But sometimes you wouldn't?
A. Probably I would.
Q. You probably would know?
A. Yeah.
Q. How come there are times when you don't think that you would know?
A. Maybe you know what it said, but you never heard of the word before.
Q. And sometimes that happens?
A. Sometimes.
Q. OK. But not very often?
A. No, not very often.
Q. Better to sound out word you don't know or ask somebody what it says?
A. Sound out word.
Q. Why?
A. Because then you learn.
Q. Whole sentence that you couldn't understand.
A. I would try to find out what it said.
Q. Anything else that you could do? How would you try to find out what it said?
A. See if there is any words in there that you know, what they meant.
Q. Anything you could do to try and find out?
A. Mm, no.
Q. No? OK.

This third-grade good reader mentioned the same decoding strategies as the poor reader, but also recognized that occasionally a word might be decoded that would be a "new word" to the child and hence he would not know the meaning. Sounding out rather than asking for a word to be decoded for you was argued to be advantageous for learning. In addition, the good reader had an idea of how to begin decoding a sentence.

By the grade-six level, children were more confident in their replies. For example, one *sixth-grade poor reader* responded as follows:

Q. Reading and come to a word that you don't know.
A. I just spell it out. [Note: Spelling was being taught using a phonics method.]
Q. Anything else?
A. Put your hand up and ask teacher.
Q. Anything else?
A. That's all I can think of.
Q. Difference between knowing what a word says and what it means. If you knew what it said, would you know what it meant?
A. No. Because you just found out what the word was. Probably wouldn't know the meaning.
Q. Better to sound out word you don't know or ask somebody what it says?

A. Sound it out. Because then you won't be bugging anybody else.
Q. Whole sentence that you couldn't understand.
A. I'd ask the teacher.
Q. Anything else?
A. Ask one of my friends.

The same two decoding strategies as just seen in the third-grade responses (sounding out and asking) were suggested as word attack strategies. In addition, by sixth grade the reader realized that many words could be decoded that would be "new words" and, consequently, he would not know what they meant. It also is interesting to note that this sixth-grade poor reader echoed the sentiments of the previously cited third-grade poor reader when asked whether it was better to sound out a word or ask somebody what it says. However, when it was a sentence rather than a word that was in question, "asking" was considered to be the appropriate strategy.

By the sixth-grade good reading level, responses were sophisticated and presented in an orderly, coherent fashion. For example, one *sixth-grade good reader* responded as follows:

Q. Reading and come to a word that you don't know.
A. Sometimes I sound it out or, if I can't get that, I look it up in the dictionary and it has, uh, the way you pronounce it in there and, um, then that's all I do.
Q. Difference between knowing what a word says and what it means. If you knew what it said, would you know what it meant?
A. No, I think if you know how to say it, you just know how to pronounce it, but you don't know the meaning of what it, what it means, and—
Q. Mm-hm. Why might you not know?
A. Because, uh, well first you didn't know the word in the first place, and then you sounded it out. You can say it, but you don't know what it means.
Q. Better to sound out word you don't know or ask somebody what it says?
A. To sound out a word that you don't know.
Q. Why?
A. Because if you ask somebody, all you're doing, uh, you're not helping yourself. You get more practice if you try to learn yourself.
Q. Whole sentence that you couldn't understand.
A. Well, I'd try to sound out all of them and if I couldn't get them, I'd look them up in the dictionary.
Q. Anything else?
A. That's about it.

In addition to the decoding strategies mentioned by the previous children, this sixth-grade good reader suggested using a dictionary, which would provide both pronunciation and meaning. On the whole, answers were more complete and more sophisticated.

In general, young and, to a certain extent, poor readers gave little indication that decoding skills were understood. Instead, the answers indicated that younger/poorer readers had acquired *coping,* rather than metacognitive, skills. For example, several younger/poorer readers indicated that they would sound out a word that they did not know because "the teacher wouldn't tell me anyway" or "the teacher doesn't like it when you talk to your friends." In addition, many younger/poorer readers had little idea of what to do with a sentence that they did not understand. In contrast, the older/better reader tended to combine strategies into a "plan of attack."

It appears, overall, that the younger and, to a lesser extent, the poorer readers have limited decoding skills. This is not to say that they know nothing of the primary skills of reading. There is evidence to suggest that they have gathered some information from what they have been taught. Unfortunately, in the case of the poorer reader, the rationale given for the use of a good decoding strategy is one that indicates that the child has learned to *cope* (in the sense of dealing with the teacher) rather than learned why or how a particular strategy is useful.

The decoding verbalization items also were used to compute a single score for each child that would indicate that child's *overall* level of achievement on decoding verbalization. The computed scores are described in the General Method section. These computed scores on decoding verbalization, presented in Table 7-1 in a later chapter, increased with grade but not with reading ability (Appendix E).

Relationship Between Performance and Verbalization

In order to examine the overall relationship between performance and verbalization, the two computed scores for decoding and the nonverbal IQ scores were entered into a stepwise multiple-regression equation, with scores on Gates-MacGinitie reading comprehension as the criterion variable. One of the purposes here was to see which was the better predictor of reading level, the performance items or the interview data.

At third-grade level, the multiple-regression equation accounted for 59.2% of the variance ($R = .7691$). The equation indicated that decoding performance accounting for 54.5% of the variance, nonverbal IQ accounted for an additional 2.2% of the variance, and decoding verbalization accounted for an additional 2.4% of the variance. At the sixth-grade level, the multiple-regression equation accounted for 66.0% of the variance ($R = .8121$). Decoding performance accounted for 55.7% of the variance, nonverbal IQ accounted for an additional 9.9% of the variance, and decoding verbalization accounted for only an additional 0.4% of the variance. In a predictive sense, then, decoding performance was a far more useful measure than decoding verbalization. It also should be noted that the nonverbal IQ measure accounted for relatively little additional variance above the computed performance scores.

Table 3-4. Decoding: Numbers of Children Categorized as LOW, MIMIC, TRANSITION, and META

	Grade 3			Grade 6			
Category	Poor	Average	Good	Poor	Average	Good	%
LOW	19	14	6	4	1	0	30.6
MIMIC	5	5	5	7	3	0	17.4
TRANS.	0	4	9	3	7	5	19.4
META	0	1	4	10	13	19	32.6

Metacognitive Categorization

As one way to take yet another view of the data, the children were classified as LOW, MIMIC, TRANSITION, or META according to the definition of meta described in the General Method section. The frequency of children falling into each category at each reading level is presented in Table 3-4. The overall chi-square analysis of these frequencies was significant (see Appendix D).

As can be seen in Table 3-4, there is a progression from being low on both performance and verbalization to being mature metacognizers. The third-grade poor readers were classified almost exclusively as LOW (19 out of 24), whereas the sixth-grade good readers were classified almost exclusively as META (19 out of 24). Very few children above the third-grade average reading level were classified as LOW, and very few children below the sixth-grade level were classified as META. In addition, the majority of children were classified as either LOW or META, with relatively few falling in the MIMIC or TRANSITION categories. Of those classified as MIMICS, most children came from the sixth-grade poor reading level or lower. Comparing the number of children in less mature categories (LOW, MIMIC, and TRANSITION) with the number of children in the META category, the number of mature metacognizers increased significantly both with grade and with reading ability (see Appendix D).

General Discussion of Decoding Analyses

Apparently, it is not reasonable to assume that poor readers, even after three to six years of teaching, have a good grasp of the primary reading skills of decoding. The third-grade poor readers seem to have difficulty in both decoding performance and decoding verbalization tasks. It appears that the children have had a great deal of difficulty decoding and have little idea of what decoding is all about. In addition, it appears that young readers (third grade) in general have a certain amount of difficulty decoding printed symbols, and also have a somewhat limited understanding of how to go about the task of decoding.

In summary, the following conclusions seem reasonable from the decoding analyses. (1) As grade and reading ability increase, so does decoding performance. That is, older children tend to be better readers (but not always), and better readers tend to be better decoders (but not always). The relationship holds even with the variance attributed to nonverbal IQ controlled. Thus, it seems insufficient just to argue that better readers are just "smarter." (2) The ability to make appropriate verbalizations about decoding increases with grade, and on some interview items with reading ability. However, on most items the reading ability effect is not strong. Among poor readers, there also appears to be a tendency to mimic (a coping strategy). We infer from this that older readers 'know' more about decoding in a metacognitive sense, and that better readers know more about certain aspects of decoding. We further infer, from the mimicking, that children who have trouble with reading nevertheless learn to say some of the 'right things.' (3) Decoding performance is a much more useful predictor than decoding verbalization. Both decoding verbalization and nonverbal IQ add relatively little predictive power at either grade. From this we infer that if you want to predict global reading scores, the best approach is to use a performance test rather than talk to children about what they know. (4) In a strict senes (taking into account both performance and verbalization), the number of mature metacognizers increases with grade and with reading ability. From this we infer that older/better readers know more about what they are doing in a metacognitive sense.

CHAPTER 4

Comprehension and Strategies

As we discussed in our introductory chapter, this study examined the development of cognitive and metacognitive aspects of decoding, comprehension, and advanced strategies, and of language, attention, and memory, all in regard to level of reading ability. The results for comprehension and strategies (i.e., Sets 3, 4, 5, and 6 in Table 2-1) are reported in this chapter. A brief review of the relevant literature follows.

Reading Comprehension: Cognition

As many of our readers probably realize, one of the problems in reviewing the literature on reading comprehension is the problem of defining "comprehension" (e.g., Carroll, 1972). Researchers have varied to a great extent in their approach to reading comprehension (e.g., story grammers as in Thorndyke, 1977, good vs. poor readers, interactive models, etc.). Due to the magnitude of the task of providing even an "adequate" review of reading comprehension research, the reader is referred to the following sources for additional information: Adams & Bruce, 1980; Furth, 1978; Gibson & Levin, 1975; Golinkoff, 1976; Grueneich & Trabasso, 1979; Murray & Pikulski, 1978; Otto & White, 1982; Pichert, 1979; Reder, 1978; Royer, Hastings, & Hook, 1979; Smith, 1971; Spiro, Bruce, & Brewer, 1980; Stanovich & West, 1979; Trabasso, 1980.

Reading Comprehension: Metacognition

Unfortunately, we know very little about metacomprehension, or the development of metacomprehension processes, particularly with regard to reading. However, many researchers have hypothesized the importance of metacognitive aspects of reading (e.g., Baker & Brown, in press; Brown, 1980; Forrest-Pressley & Gillies, 1983; Yussen, Matthews & Hiebert, 1982) and recent research has begun to shed some light on the area. For example, Markman (1979) asked children to listen to stories in which logical inconsistencies had been embedded. Using a probe procedure, she then determined whether or not the inconsistencies had been detected. She found that younger children (third grade) were less likely to notice inconsistencies when listening to passages and needed considerable prompting before errors in comprehension were detected. In addition, Baker (1979) found that college students reported confusions involving details more frequently than those involvng main points. Retrospective reports indicated that failure to report confusions was often due to "repair" strategies (i.e., students rationalized the inconsistency) rather than comprehension-monitoring failures.

Fortunately, there probably are ways in which we can encourage children's use of monitoring. For example, Markman and Gorin (1981) asked children, ages eight and ten years, to listen to stories containing one of two types of errors, that is, inconsistencies (such as those used by Markman, 1977, 1979) or falsehoods (i.e., statements that would be recognized as wrong if evaluated against common knowledge of the world). Telling children what type of error to look for increased the probability that errors would be detected, although inconsistencies were easier to detect than falsehoods.

Certainly all available evidence points to a developmental trend in knowledge of comprehension skills, even though there are still many unanswered problems in the area. One major problem is that studies of metacomprehension have not related these "meta" measures to measures of reading ability, per se. Also, the child's awareness of comprehension processes (as measured by self reports) has been more or less ignored. Considerable research is needed before the pattern of development of metacognitive aspects of reading comprehension can be defined clearly.

Reading Strategies: Cognition

As we suggested in the introduction, we would like to think that reading is more than just decoding symbols. In fact, when considering the mature reader, decoding seems to be a relatively small part of the process. Rather, the mature reader seems to be able to decode "automatically," and cognitive efforts are focused on extracting the information that will be meaningful within the context of the purpose of reading the material. These cognitive activities, originating from the reader's control of print, have been referred to by Rothkopf (1972) as "mathemagenic activities," that is, activities that give rise to the learning of text and are relevant to the achievement of the goal at hand.

According to Rothkopf and Billington (1975), reading performance is affected by the experience of the reader, the things that one does to process the information read, and the factors related to the task conditions. For example, when goals are provided, the reader seems to be able to extract and remember specific information from a text with little incidental learning. However, in the absence of goals, the information that is remembered is extremely variable. For the most part, however, these studies of mathemagenic activities have used adults as subjects. It is important to know if these skills are present during childhood and when the beginning reader first shows signs of use of these advanced skills.

Unfortunately, very little research has been done that would indicate the pattern of development of advanced reading strategies. Indeed, the little research that is available often concentrates on a very narrow range of ages. For example, several studies have reported that the provision of advance organizers to help students relate the material to be read to what is already known does not seem to aid either third- or sixth-graders (Clawson & Barnes, 1973; Proger, Carter, Mann, Taylor, Bayuk, Morris & Reckless, 1973). However, this type of advance knowledge does seem to aid ninth-grade students, especially those of low ability, and especially when they were forced to write a summary of the advance organizer (Allen, 1970). When cues take the form of inserted questions, Swensen and Kulhavy (1974) found that fifth- and sixth-graders did benefit from the added information. In addition, Forrest and Barron (1977) found that only sixth-graders (in comparison with children from grades two and four) could adjust their reading to meet goal requirements. Nevertheless, more research is needed to outline the manner in which reading strategies are acquired and the extent to which they are used in a normal classroom situation. We will not devote additional attention to reading strategies except to note that extensive reviews exist for the use of strategies, particularly in adults (e.g., Anderson, 1980; Anderson & Armbruster, 1980; Anderson & Biddle, 1975; Ausubel, 1978; Barnes & Clawson, 1975; Ford, 1981; Frase, 1972, 1975, 1977; Levin & Pressley, 1981; Rothkopf, 1972; Rothkopf & Billington, 1975; Tierney & Cunningham, 1980; Walker & Meyer, 1980).

Reading Strategies: Metacognition

The little that we do know about the development of reading strategies does not seem to be incongruent with what is known about metacognition. Rothkopf and Billington (1975) concluded that reading performance is due to the experience of the person (what the subject knows about words and concepts), the processing effort (the mathemagenic activities), and the characteristics of the text. Further, according to Brown (1980), the activities that the mature reader engages in include: (1) clarifying the purposes of reading (i.e., understanding the task demands), (2) identifying the important aspects of the message, (3) allocating attention to relevant information, (4) monitoring activities to determine if comprehension is occurring, (5) engaging in review and self-testing, (6) taking corrective measures if necessary, and (7) recovering from disruptions and distractions. Brown then argued that the efficient reader learns to evaluate strategy selection in terms of the situation (i.e., the goal demands).

We know of very little research investigating the development of metacognitive aspects of reading strategies. However, Myers and Paris (1978) found in an interview study (following the Kreutzer, Leonard & Flavell, 1975 model) that older readers were more aware of the factors affecting their reading than younger readers. Unfortunately, strategies of reading were not examined specifically in that study. Much systematic research is needed before the development of metacognitive aspects of reading strategies can be clearly understood.

For the purpose of this investigation, we will consider *mature* reading as a set of various skills that leads to a high level of comprehension with the expenditure of the least amount of effort. Several writers (e.g., Gibson, 1972; Gibson & Levin, 1975; Brown, 1980) have suggested that one characteristic of a mature reader is the ability to adjust reading activities to meet the purposes of any given reading situation (i.e., to adopt a reading strategy appropriate to the reading task). It also has been suggested (e.g., Brown, 1980) that success in any given reading situation depends not only on the flexible *use* of reading skills, but also on the ability to *monitor the progress* of reading so that failures of comprehension can be corrected. In addition, we argue that the ability to monitor comprehension depends upon what a reader knows about his or her own comprehension processes.

While comprehension has been of considerable continuing interest, we know of no researchers who have examined the development of metacomprehension *in conjunction with* the development of reading. We are using the term "comprehension" here as it normally has been defined; that is, information is comprehended by the reader to the extent to which it can be used, recalled, or recognized. In contrast, metacognitive aspects of comprehension involve knowing when you have understood what you have read, knowing what you do not understand, and being able to use this knowledge to monitor comprehension.

In addition, the mature reader must be flexible in his reading skills (i.e., have the ability to use strategies of reading for a purpose). This ability involves the use of skills such as skimming and reviewing in order to extract information to meet a specific purpose. The metacognitive aspects of these advanced strategies involve knowing that you read differently in different situations, that there are things you can do in order to aid retention, and that some methods are more appropriate and efficient than others in any particular situation.

When we begin the task of operationalizing the various ideas involved in mature reading, a number of types of measures are possible. Comprehension, in a cognitive sense, can be measured by the degree to which a reader can use the information that has been read (e.g., by giving a standard test after a passage has been read or by asking the reader to compose a title for a passage). Metacomprehension can be measured by seeing how the reader subsequently makes use of the information read, is able to predict the extent to which performance has been successful, and can explain his or her knowledge of comprehension processes.

To determine cognitive ability, advanced strategies can be indexed by the degree to which a reader actually reads differently in different reading situations. This can be assessed simply by comparing comprehension measures taken after the reader has read for different purposes. For example, a person should score higher on a comprehension test if he has *studied* a reading passage than if the passage has been *skimmed* for one specific piece of information. And to examine metacognition, knowledge about advanced reading strategies can be assessed in an interview in which a reader is asked about different ways to read in different situations.

An alternative, and perhaps less direct, way to approach reading strategies is in the context of reading flexibility, which should result in reading efficiency. Presumably, a flexible reader, one who adjusts his or her reading strategies to the demands of the reading situation, will be efficient. In contrast, an unskilled or inflexible reader either will read the same "way" in all situations, or will not match an appropriate strategy to a particular situation. Either way, the end result should be low efficiency. Reading efficiency can be indexed simply by dividing a reading comprehension score by a unit of time (as in Faw and Waller, 1977). The advantage of an efficiency score is that it allows comprehension scores to be adjusted by a reading time score so that a slow reader who, for example, correctly answers 80 percent of a set of comprehension questions would receive a lower efficiency score than a fast reader who achieves the same level of comprehension but in less time. A metacognitive counterpart would be the level of efficiency achieved plus knowledge about strategies of reading for a purpose. If a reader knows about various strategies, and knows how and when to use them, then in effect, that reader in some sense knows how to be efficient.

Before proceeding further in this chapter, we should warn the reader that the structure of this chapter differs slightly from the standard structure used in some of the other chapters of this book. Since the performance measures for both comprehension and strategies are from the "reading experiment," we will present these together. Then we will complete the comprehension section (verbalization measures, regressions, metacategorizations) and follow this by completing the strategies section. At the end of the strategies verbalization section, we will present and discuss some of the qualitative changes evident in both the comprehension and strategies interviews.

Comprehension and Strategies Performance: The Children's Task

The first of the children's tasks for this part of the study was designed to examine their reading ability under different conditions. We asked each child to read a set of passages in each of four different instructional conditions. After each passage, some of which were easy and some of which were hard, the child answered a comprehension test composed of two types of items, some based on thematically important parts of the passage and some based on thematically less important parts. After responding to each question, the child was asked to indicate whether he or she was "sure" or "not sure" that the chosen answer was correct. We used these reponses to generate prediction-accuracy scores.

Obviously, we have implied that there were various conditions and procedures involved in this "reading study," and we will describe these in more detail below. As to how these comprehension scores taken in this reading study reflect the Sets presented in Table 2-1, (1) differences in performance under the various instructional conditions are relevant to strategies (Set 5, Table 2-1), and (2) the prediction-accuracy scores provided part of the data relevant to evaluating metacomprehension (Set 4, Table 2-1). In addition, during a separate session we asked each child a series of standard interview questions about reading. The procedure and results for the interviews (relevant to Sets 4 and 6, Table 2-1) are described in a later section of this chapter.

Instructions to the Children

We asked each child to read passages that were presented under four different instructional conditions: "read for fun," "read for a title," "read to find one specific piece of information (skim)," and "read to study." In the *Fun* condition, we asked the children to decide whether or not the story was a good story (i.e., whether other boys and girls their age would enjoy reading the story.) We told them that as soon as they had finished reading the story, they would be asked to give the story a "mark" out of 10, with 10 meaning the story was really good and 1 meaning they did not like the story at all. We also indicated that nothing else (e.g., reading time, comprehension questions) was important. In the *Title* condition, we asked the children to think of the best possible title for the story. We told them that they would be asked for their title as soon as they had completed the story, but that nothing else (e.g., reading time, comprehension questions) was important. (The titles subsequently were judged by adult raters for their appropriateness to the theme of the story.) In the *Skim* condition, we asked the children to finish with the story as quickly as possible and that it was important that they find the answer to one "special question." The special question was usually the name of a person, place, or thing, and occurred on the first page of the story. The children were told what the question was immediately before starting to read the story and were asked to repeat the question to the experimenter to ensure that they knew what information they needed to find. We told the children that they would be asked for the answer to the special question as soon as they indicated that they were done with the story. We also indicated that nothing else (e.g., comprehension questions) was important. In the *Study* condition, we told the children that the comprehension questions were important and that they should try to get as many as possible correct. We also indicated that nothing else (e.g., reading time) was important. Analyses of title appropriateness, skimming questions, and "fun" ratings are presented in Appendix F. Complete instructions to the children are presented in Appendix G.

Difficulty of the Passages

According to Gunning (1968), the optimal level for a child to read is the level immediately below the child's current grade level. Since the difficulty of a reading passage is an important variable in developmental research, we decided to use two difficulty levels for each grade. The approximate grade levels were 2.0 to 3.9 and 4.0 to 5.9 for third-grade easy and hard passages respectively, and 6.0 to 7.9 and 8.0 to 9.9 for sixth-grade easy and hard passages respectively. We used the Fog Readability Index (Gunning, 1968) to determine the readability level of the passages. In addition, we selected the books from which the passages were taken on the basis of their recommended reading level (as provided by the publishers). Also, as the children were asked to give the stories a "fun rating" on a scale of 1 to 10 in the Fun instructional condition, these "fun" ratings were used to verify further that the passages were appropriate to the ages of the children. The mean Fun ratings (reported in full in Appendix F) were rather high, indicating at least that the children perceived the stories as "fun." Sample passages and questions are presented in Appendix H.

Thematic Importance of Sentences

We selected excerpts (of approximately 500 words) from each of four easy stories and four hard stories. Since it is possible that retention of material may be a result of the importance of any particular unit of text to the theme of the story (Brown & Smiley, 1977), and that the importance of a unit may be relevant to the purpose of reading, we gave these excerpts to ten adults and asked them to rate sentences in the passages as important or less important to the theme of the story. We then calculated the number of people choosing each sentence as important. We classified those sentences identified as important by seven or more people as "Important Sentences," and those sentences identified by two or fewer people as important as "Less Important Sentences." We chose seven sentences of each type from each passage as the basis for a question to be included on the test given after the passage was read by the children.

The Statistical Design of the Reading Study

The overall statistical plan was a factorial design that included the variables of grade, reading ability, instructional condition, difficulty level, and importance of sentence. The first two variables are group, or between-subject, variables, while the others are repeated measure, or within-subject, variables. That is, each child read eight passages (one in each instruction by difficulty combination) and each passage contained both important and less important sentences. Within the design, each instructional condition appeared the same number of times in the first, second, third, and fourth positions, and each passage appeared the same number of times with each instructional condition. Half of the children did the easy passages first and half did the hard passages first.

Measure of the Children's Comprehension

After each passage, each child answered 14 multiple-choice questions. These test questions were based on the seven important and seven less important sentences that we described above. Each sentence was split approximately in half, and the second half was used as one of four alternatives in a multiple-choice question. We presented the multiple-choice questions in the order in which the sentences appeared in the passage. Sample questions are shown in Appendix H.

Measure of the Children's Prediction Accuracy

We created an index of comprehension monitoring based on a child's prediction of his or her own accuracy on each item of the comprehension test. After each test item, we asked the children if they were sure or not sure of the answers they had given. We assigned a score of two points if the child was accurate in a prediction (i.e., if the answer was in fact correct and the child was "sure," or if the answer was *not* correct and the child was "not sure" of it). If the child indicated uncertainty when the answer was correct, we assigned a score of one point. However, if the child was inaccurate in a prediction (i.e., an indication of "sure" was given when a response was incorrect), we assigned a score of zero. (Additional information regarding knowledge of comprehension monitoring was provided by the interview data and is presented in a later section of this chapter.)

What the Children Did

All of the reading tests were done in two sessions, each session covering both difficulty levels of two of the instructional conditions. The children were told the instructions for reading before beginning to read each story. In every case, we asked children to repeat the instructions in order to insure that the instructions had been understood. Also, we reminded them of the instructions immediately before they began to read. Immediately after each story, we asked the child for the information considered "important" for the instructional condition. Then the child completed the comprehension and prediction-accuracy tests. We also recorded the length of time the child spent reading each story. After completing the comprehension questions, we asked any incidental questions (e.g., "special" question in the Skim, Fun, Title, and Study conditions). Extra measures (e.g., fun ratings, titles, reading time) were recorded when appropriate. For those interested, the method and results for these extra measures are presented in Appendix F. In addition, we asked the third-grade good readers and sixth-grade poor readers to participate in a third reading session in which they read passages for a "controlled passages" comparison. More information can be found about these measures in Appendix I. Statistical source tables for all analyses of variance and analyses of covariance can be found in Appendix J. Significant effects of all analyses (analyses of variance, chi squares, correlations, metacategorizations) are presented in Appendix K.

Comprehension and Strategies Performance: Results and Discussion

Each child answered 14 questions on each of eight passages. Each answer was scored correct or incorrect and a total correct score was computed for each of the seven important and seven less important sentences. We analyzed the scores using a mixed analysis of variance procedure, with the factors being grade, reading ability, instructional condition, difficulty level, and importance of sentence. All subsequent individual comparisons were based on the Newman-Keuls procedure at a probability level of .05.

In general, the comprehension scores increased with grade and with reading ability, and were affected by instructional condition. In addition, comprehension scores decreased as difficulty level increased and were higher for important sentences than for less important sentences. The variables of grade, reading ability, instructional condition, and difficulty level also interacted. When reading easy passages, third-grade good readers retained more information in the Study condition than in the Skim condition, and when reading hard passages, they retained more in the Fun than in the Skim condition. Performance was higher in the Study than the Skim condition for the sixth-grade poor readers when reading hard passages, and for sixth-grade good readers when reading easy passages.

Since this interaction indicated, as expected, that the pattern of responding changed with grade and reading ability, we decided to analyze the data from each group (e.g., third-grade poor readers) separately. Each of these analyses was a repeated-measures analysis of variance, with the factors being level and importance of sentence. We were particularly interested in the three-way interaction involving grade, reading ability, and instructions. This interaction is shown in Figure 4-1 and is a primary focus for the separate analyses.

For sixth-grade good readers, performance was better on important sentences than on less important sentences. In addition, the main effect of instructional condition was significant. As shown in Figure 4-1, performance was lower in the Skim condition than in all of the other instructional conditions. These differences indicate that sixth-grade good readers adjusted their reading to meet the demands of the situation (i.e., they employed strategies of reading for a purpose).

For sixth-grade average readers, performance was better for important sentences than for less important sentences, and better for easy passages than for

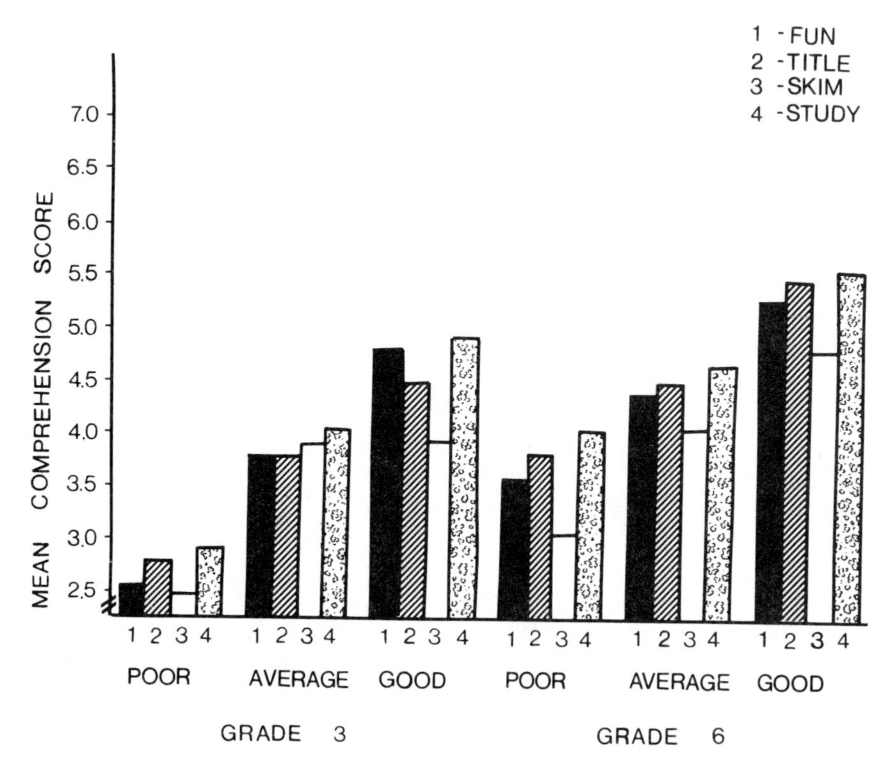

Figure 4-1. Effect of grade, reading level, and instructional condition on comprehension scores.

hard passages. At this reading level, there was no effect of instructional condition, although a pattern similar to that of sixth-grade good readers is evident.

For sixth-grade poor readers, performance was better for important sentences than for less important sentences, and better for the easy stories than for the hard stories. In addition, the main effect of instructional condition was significant. As shown in Figure 4-1, performance in the Skim condition was lower than in all other conditions, particulary the Study condition.

For third-grade good readers, performance was better for important sentences than for less important sentences. In addition, the main effect of instructional condition and the interaction of instructional condition, importance of sentence, and difficulty level was significant. Children responded correctly to important sentences at the same level, regardless of instructional condition or difficulty of passage. However, the pattern of responding was different on the less important sentences. Considering first the easy passages, performance on the less important sentences was better in the Study than in all other instructional conditions. Concerning the hard passages, performance on the less important sentences was higher in the Fun than in the Skim and Title conditions, and higher in the Study than in the Skim condition. Collapsed over difficulty and importance of sentence, performance was lower in the Skim than in either the Fun or the Study conditions.

For third-grade poor and average readers, the only significant effect was importance of sentence. Even at these reading levels, performance was higher on important sentences than on less important sentences. The effect of instructional condition was not significant, indicating that readers at these levels were not adjusting their reading to meet the demands of the situation.

In addition to the four-way interaction just described, the variables of grade and importance of sentence and the variables of grade, reading ability, and difficulty level interacted with one another. The difference between performance on important and unimportant sentences was greater for third-grade than for sixth-grade children. In addition, there was no difference in performance on easy and hard passages for all third-grade reading levels and for sixth-grade good readers. Performance on easy passages was greater for the sixth grade poor and average readers.

Moreover, the data also were analyzed using analyses of covariance, with nonverbal IQ scores as the covariate. These analyses showed a similar pattern of results. Comprehension increased with grade and with reading ability, and the interaction of grade, reading ability, difficulty level, and instructional condition was significant. Thus, it does not appear reasonable to suggest that the pattern of results is due to variability in nonverbal IQ, at least as it was measured in this study.

Finally, we combined the raw comprehension measures as described in the General Method chapter to derive a computed score that respresented each child's overall level of achievement on comprehension performance. These computed scores, presented in Table 7-1 in a later chapter, increased with grade and with reading ability.

Overall, then, comprehension scores increased with grade and with reading ability, were higher for thematically important sentences than for less important sentences, and were lower on harder passages. The pattern of comprehension scores over the four instructional conditions indicated that the ability to adjust reading strategies did change with age and with reading ability. The first change occurred with third-grade good readers whose comprehension scores dropped when they were asked to find one specific piece of information (i.e., in the Skim condition compared to the Study and Fun conditions). At the sixth-grade level, performance in the Study condition also was higher than in the Skim condition (although not significantly for average readers). In addition, it was only with sixth-grade good readers that performance was significantly higher in *all* other instructional conditions than in the Skim condition. In effect, with both increasing grade and increasing reading ability, there were corresponding increases both on reading comprehension per se, and in the tendency to read strategically (i.e., to respond to the instructional demands of the situation).

Table 4-1. Prediction-Accuracy Means: Grade and Reading Ability

	Poor	Average	Good	Mean
Grade 3	8.15	9.36	10.54	9.35
Grade 6	10.09	11.23	11.64	10.99
Mean	9.12	10.30	11.09	

Prediction-Accuracy Scores

As we explained earlier, after choosing a response for each multiple-coice question, each child was asked if he or she was sure or not sure of the answer given. These judgments were scored according to their accuracy of prediction and analyzed in a mixed analysis of variance, with the variables of reading ability, instructional condition, difficulty level, and importance of sentence.

The ability to predict the correctness of the response increased with grade and with reading ability. The appropriate means are shown in Table 4-1. Good readers were better at predicting accuracy than average readers and average readers, were better at predicting accuracy than poor readers.

It also was easier to predict the accuracy of answers to questions involving important sentences than for those involving less important sentences. In addition, accuracy of prediction varied with instruction, being lower in the Skim condition than in any other instructional condition. Also, instructional condition and importance of sentence interacted with each other. Important sentences, for example, were predicted less accurately in the Skim than in all other instructional conditions. However, less important sentences were predicted less accurately in the Skim than in the Study condition, in the Skim than in the Fun condition, and in the Title than in the Study condition. Difficulty level also interacted with grade and with reading ability. It appeared that the ability to predict accurately decreased with difficulty at the third-grade good reading level, but difficulty made no difference at the remaining levels. Because it did not appear as if the pattern of responding was changing systematically with grade and with reading ability, we decided not to analyze the data further, as was done with the comprehension scores.

In summary, analyses of the prediction-accuracy scores showed that success at prediction increased with grade and with reading ability. Although success of prediction varied with instruction (i.e., was lower in the Skim condition), the effect of instructions did not interact with age or with reading ability. Thematically important sentences also were predicted more accurately than less important sentences. We feel, then, that the younger/poorer reader not only has a lower level of comprehension, but also is less able to monitor comprehension than the older/better reader, who is more successful at predicting accuracy of comprehension. Therefore, it is quite possible that the younger/poorer reader does not realize when a comprehension problem occurs. Finally, these prediction-accuracy data are assumed to reflect metacomprehension, and we will discuss them again with the results of the interview data on knowledge about comprehension.

Comprehension Verbalization

Each child also participated in a structured interview and responded to a series of questions in various categories. Interview items relevant to comprehension (Set 4, Table 2-1) are reported here. Each interview item and the scoring key are presented in full in Appendix L. For each item, the number of children in each grade and reading level giving any particular level of response for the item is shown in Table 4-2 along with the appropriate overall chi-square and probability level.

We asked each child the following questions in order to assess the child's knowledge of his or her ability to assess comprehension and the child's knowledge of the comprehension-monitoring process.

Question C-V-1: How do you know when you are ready to write a test?

Approximately 40% of each of the third-grade poor and third-grade average readers could not think of any way in which they could tell when they would be ready to write a test. However, about an equal percentage could at least cite an "after reading" or a rehearsal strategy. The overall chi square was significant, indicating a relationship between performance and grade level, reading ability, or both. The subsequent analyses indicated that the ability to produce a mature verbalization increased with grade.

Question C-V-2: Would you know how well you had done on the test before you got it back? How?

By far the most common answer of the younger/poorer reader was that you "can't tell before you get the test back." As grade and reading ability increased, so did the tendency to cite cues such as the length of time to finish, the number that you are sure of, etc., which indicate the level of difficulty of the test. The overall chi square was significant, and subsequent analyses showed that the item differentiated mainly between grades.

Question C-V-3: How would you know when you knew enough about a game to be able to teach someone else about it?

Similarly, on this item, older children gave more sophisticated responses.

In summary, it appears that the ability to make sophisticated verbalizations about how one monitors one's own comprehension is a function of grade considerably more than a function of reading ability. However, in spite of the fact that the statistical tests of significance do not indicate a strong effect of reading ability, the reader should note that for each item there is a gradual but consistent increase in level of responding as a function of reading ability.

Table 4-2. Comprehension Verbalization: Frequency of Scores

		C-V-1	C-V-2	C-V-3
Grade 3				
Poor	0	9	17	12
	1	11	4	5
	2	4	3	7
Average	0	10	16	11
	1	9	4	5
	2	5	4	8
Good	0	4	11	6
	1	14	5	8
	2	6	8	10
Grade 6				
Poor	0	6	6	1
	1	10	4	6
	2	8	14	17
Average	0	3	2	2
	1	13	8	4
	2	8	14	18
Good	0	2	0	2
	1	9	7	7
	2	13	17	15
Chi square		18.41*	49.34***	31.64**

*$p < .05$ **$p < .001$ ***$p < .0001$

We also assessed the relationship between each verbalization item and reading ability at each grade level by means of correlations and partial correlations, controlling for nonverbal IQ. These correlations are reported in the statistical analysis appendix for comprehension and strategies (Appendix K). As you will see, some items correlated with reading ability at each grade, but when the variance attributed to nonverbal IQ was removed, the relationship became weaker.

We also combined the comprehension verbalization scores and the prediction-accuracy scores as described in the General Method section to derive a computed score that would indicate each child's knowledge and monitoring of comprehension processes. This overall measure, presented in Table 7-1 in a later chapter, increased with grade and with reading ability.

Comprehension: Relationship Between Performance and Verbalization

We used the computed scores on comprehension performance and on comprehension verbalization in stepwise multiple regressions to predict reading ability at each grade level. Nonverbal IQ also was included in the regression analyses.

Table 4-3. Comprehension: Numbers of Children Categorized as LOW, MIMIC, TRANSITION, and META

	Grade 3			Grade 6			
Category	Poor	Average	Good	Poor	Average	Good	%
LOW	19	11	5	12	1	1	34.0
MIMIC	1	1	2	5	8	2	13.2
TRANS.	2	3	3	2	0	1	7.6
META	2	9	14	5	15	20	45.2

At the third-grade level, the multiple-regression equation accounted for 38.57% of the variance in reading, as assessed by the Gates-MacGinitie Reading Comprehension Test (R = .6210). The equation indicated that comprehension performance accounted for 36.99% of the variance, nonverbal IQ accounted for an additional 1.50% of the variance, while comprehension verbalization accounted for only an additional 0.09% of the variance. At the sixth-grade level, the multiple-regression equation accounted for 58.52% of the variance (R = .7650). Comprehension performance accounted for 44.91% of the variance, nonverbal IQ accounted for an additional 13.53% of the variance, while comprehension verbalization accounted for only an additional .08% of the variance. In a predictive sense, then, we feel that it is quite clear from the measures used here that comprehension performance is a more useful measure of reading ability than either nonverbal IQ or comprehension verbalization.

Comprehension: Metacognitive Categorizations

We also classified the children according to the definition of "meta" that we described in the General Method chapter. The number of children falling into each category at each grade and reading level is presented in Table 4-3. The overall chi-square analysis of these frequencies was significant (and is presented in Appendix K).

As the reader can see, there is a progression from being low on both performance and verbalization to being mature metacognizers. In this case, the progression appears to be strongly related to reading ability as well as to grade level. A majority of the third-grade good readers and sixth-grade average and good readers were classified as META, whereas third-grade poor and average readers and sixth-grade poor readers were classified predominately as LOW. In addition, a relatively low percentage of children from any category fell in either the MIMIC or TRANSITION categories (approximately 20 percent). When we compare the number of children in less mature categories with the number of children in the META category, the number of mature metacognizers increased with grade and with reading ability. Poor and average readers differed significantly

Table 4-4. Strategies Verbalization: Frequency of Scores

		S-V-1	S-V-2	S-V-3	S-V-4	S-V-5	S-V-6
Grade 3							
Poor	0	10	15	4	2	6	8
	1	12	9	19	21	18	9
	2	2	0	1	1	0	7
Average	0	3	13	3	3	4	6
	1	20	10	18	20	20	5
	2	1	1	3	1	0	13
Good	0	6	13	0	2	3	0
	1	16	9	19	17	18	9
	2	2	2	5	4	2	14
Grade 6							
Poor	0	2	9	0	3	3	0
	1	16	13	13	13	16	2
	2	6	2	11	6	3	20
Average	0	2	5	1	1	2	0
	1	19	10	12	17	15	1
	2	3	9	11	6	7	23
Good	0	0	5	0	1	0	0
	1	20	7	7	11	10	2
	2	4	12	17	12	14	22
Chi square		25.09*	39.40**	42.25**	22.87*	43.00**	53.99**

Note: Where cell frequencies do not sum to 24, there were small bits of missing data for individual children on specific items. These children were not included in the frequency analyses on those items.

*$p < .01$ **$p < .0001$

at both grade levels. In addition, third-grade good readers were more apt to be classified as META than sixth-grade poor readers.

Strategies Verbalization

Interview items relevant to strategies are reported here (Set 6, Table 2-1). Items and scoring criteria are reported in full in Appendix M. For each item, the number of children in each grade and reading level giving any particular level of response for the item is shown in Table 4-4 along with the appropriate overall chi-square value and probability level.

We asked the following questions to assess the child's knowledge of different reading strategies (e.g., does the child know of the different ways in which he or she can read in order to be efficient in any given reading situation).

Question S-V-1: What do you do when you read in preparation for a test?

As can be seen in Table 4-4, it appears that the older/better reader is more likely to be aware of the need for rehearsal before a test. However, comparing the number of immature responses with the number of mature responses, the differences between grades and between adjacent reading groups were not significant.

Question S-V-2: Is there anything that you can do to make what you are reading easier to remember?

Only the older/better readers tended to indicate that strategies such as taking notes, remembering main points, and other- and self-testing make it easier to remember what you have read. The overall chi square was significant, and subsequent analyses showed that the ability to think of ways to remember what you have read increased with grade and with reading ability. Poor readers differed from average readers, particularly at the sixth-grade level.

Question S-V-3: How would you find the name of a place in a story? (An attempt to assess knowledge about skimming.)

In order to find one specific piece of information, the older/better reader usually suggested a 'skim' strategy rather than a 'just read' strategy, as was often suggested by the younger/poorer reader. Subsequent analyses showed that the ability to explain a 'skim' strategy increased with grade. It is interesting to note that at times, the older/better reader was reluctant to suggest that skimming was the "best" alternative, because "the teacher always asks more than one question."

Question S-V-4: How would you remember a story so that you could tell it to a friend later?

On this item, the younger/poorer reader tended to suggest writing the whole story down, while the older/better reader suggested trying to remember only the important parts. Once again, comparing the frequencies of immature and sophisticated responses, it was found that the ability to suggest strategies for remembering a story increased with grade and with reading ability. The difference in reading ability was mainly between average and good readers.

The next item was an extention of Question S-V-4 and was an attempt to assess the extent to which children were able to predict the demands of the task, in that the importance of units can affect recall.

Question S-V-5: How much of the story would you remember?

The younger/poorer readers tended to be unrealistic in the amount of a story that they would be able to remember. As grade and reading ability increased, there was a progression from claiming "all or none" retention, to a quarter or half of the story, to an indication that only important parts (main ideas, and not word for word) would be recalled. The ability to produce realistic predictions increased with grade and with reading ability. The effect of reading ability was mainly due to a difference between average and good readers.

Question S-V-6: How would you think of a title for a story?

When thinking of a title, the older/better reader usually suggested the need for thematization. This realization seemed to be lacking in younger/poorer readers. The ability to produce a sophisicated response to this item was mainly a function of grade.

We also assessed the relationship between each verbalization item and reading ability at each grade level by means of correlations and partial correlations, controlling for nonverbal IQ. These correlations are reported in the statistical analyses appendix for comprehension and strategies (Appendix K). As the reader can see, some items correlated with reading ability at each grade but when the variance attributed to nonverbal IQ was removed, the relationships were weaker.

We also combined all of the strategies verbalization scores, as we described in the General Method section, to derive a single computed score for each child's overall level of knowledge about strategies of mature reading. This measure, presented in Table 7-1 in a later chapter, increased with grade and with reading ability.

Strategies: Relationship Between Performance and Verbalization

As with decoding and comprehension, we felt that it was desirable to compute for each child a single score that would represent performance on advanced strategies and another to represent verbalization about advanced strategies. The verbalization computed scores presented no problem and have been presented above. However, deriving a single computed score to reflect strategies performance did present a problem since the inference of use of strategies was based on a comparison of comprehension scores under different instructional conditions. On the basis of the differences of comprehension as a function of instructional condition as described above, it already has been concluded that older/better readers are more prone to engage in different reading strategies in the performance sense. The problem is to index this use of strategies with one number for the sole purpose of further calculations and treatments of the data. One solution to this problem is to think of strategies in terms of flexibility, or efficiency of reading. If the reader is able to use advanced strategies appropriately, then he or she is an efficient reader. Admittedly, efficiency represents only one aspect of use of advanced strategies, and it would not be entirely appropriate to infer strategies solely from the efficiency rating of any particular reader. But, given the subtleties of the data (e.g., the reluctance of good readers to "skim" in the strictest sense, the speed at which good readers read in any situation [Appendix F], the difference in variance among reading groups, and the use of coping

Table 4-5. Strategies: Numbers of Children Categorized as LOW, MIMIC, TRANSITION, and META

	Grade 3			Grade 6			
Category	Poor	Average	Good	Poor	Average	Good	%
LOW	20	11	8	5	3	0	32.6
MIMIC	3	2	1	12	3	1	15.3
TRANS.	1	6	7	1	2	1	12.5
META	0	5	8	6	16	22	39.6

strategies by poor readers), the use of efficiency scores as an index of advanced strategies seemed to be an acceptable alternative. Such scores were computed and henceforth are referred to as strategies (efficiency) performance computed scores. Following a procedure suggested by Faw and Waller (1977), efficiency scores were computed by dividing each child's individual comprehension scores by the number of seconds (in log 10) the child took to read that particular passage. These scores were combined, as described in the General Method section, to derive a score that represented each child's overall level of achievement on flexibility in mature reading performance. This measure, presented in Table 7- in a later chapter, increased with grade and with reading ability.

We entered the strategies performance computed score (based on efficiency scores) and the strategies verbalization computed score (based on interview questions on advanced strategies) into stepwise multiple-regression equations to predict reading ability at each grade level. Nonverbal IQ also was included in each equation. At the third-grade level, the multiple-regression equation accounted for 50.88% of the variance (R = .7133). The equation indicated that performance accounted for 50.44% of the variance, verbalization accounted for an additional 0.38% of the variance, and nonverbal IQ accounted for only an additional 0.06% of the variance. At the sixth-grade level, the multiple-regression equation accounted for 69.33% of the variance (R = .8326). The equation indicated that performance accounted for 62.90% of the variance, nonverbal IQ accounted for an additional 6.27% of the variance, and verbalization accounted for only an additional 0.16% of the variance. In a predictive sense, then, it appears that strategies performance (efficiency) is a much more useful measure of reading ability than either nonverbal IQ or strategies verbalization.

Strategies: Metacogitive Categorizations

We also classified the children according to the definition of "meta" that we described in the General Method chapter. The number of children falling into each category at each grade and reading level is presented in Table 4-5. The overall chi-square analysis of these frequencies was significant (and is reported in Appendix K).

As the reader can see, there was a progression from being low on both performance and verbalization of strategies to being mature metacognizers. The third-grade poor readers were classified almost exclusively as LOW, while the sixth-grade average and good readers tended to be classified as META. In addition, less than 30 percent of all the children fell in either the MIMIC or TRANSITION categories. The only concentration at all was with the sixth-grade poor readers, of whom 50 percent were classified as MIMICS. It might be that sixth-grade poor readers have remembered enough of what they have been taught to mimic an appropriate response. However, they seem to lack the understanding that would lead them to use strategies in an efficient manner. When we compared the number of children in less mature categories with the number of children in the META category, we found that the number of mature metacognizers increased with grade and with reading ability. The reading ability difference mainly was due to a difference between poor and average readers, particularly at the sixth-grade level.

Comprehension and Strategies: Qualitative Changes

As with decoding, many qualitative differences can be seen among children at the various levels simply by reading the interview transcripts. Selected excerpts from good and poor readers at each grade level follow. (In the interest of brevity, the questions have been abbreviated, but the children's responses are reported verbatim.)

The younger/poorer reader seemed to have difficutly thinking of ways in which he could monitor comprehension. For example, one *third-grade poor reader* responded as follows.

Q. Knowing when ready to write test.
A. Just practice reading it.
Q. Then you would be ready?
A. Yeah.
Q. Telling parents how test was.
A. I would just put, um, it was easy.
Q. How would you know?
A. Because we did it in school already.
Q. Would you know before you got it back?
A. I'd wait until teacher finished marking it.
Q. New game. Teaching friend. Could you do it?
A. No.
Q. What would you need to know?
A. I would have to know what the game is called and what do you have to do.
Q. What would you need to know before you could teach somebody?
A. How do you play it?

Q. How could you tell when you knew enough to teach it?
A. Well, you would first have to play it.
Q. As soon as you played it once, then you could teach somebody else? OK.

Even if the younger/poorer reader realized that there was a problem with comprehension, it seems unlikely that he would have any idea of what might be done to correct the situation. When asked about reading strategies, the same third-grade poor reader responded as follows.

Q. Reading for a test.
A. I would read it. I'd practice it.
Q. Read same way for fun?
A. Yeah.
Q. Anything to make what you are reading easier to remember?
A. Nothing.
Q. Story. Name of place where people lived.
A. Look through.
Q. All the way through?
A. Yeah.
Q. Would you just look through or would you read it?
A. I would just look through.
Q. That is the best way?
A. Yes. That's about it.
Q. Remembering story to tell a friend.
A. Write it on piece of paper.
Q. How much of it would you remember?
A. Half of it.
Q. Thinking of a title.
A. Just say a title and see if it makes sense, like if you are talking about animals, you would just put "Zoo Animals."
Q. So you would just kind of think of one then.
A. Yeah.
Q. How would you tell if it made sense?
A. Just see what you are talking about.

Even though the level of response given by this reader was generally low, we must admit that certain knowledge was exhibited. Even though there was no clear explanation of skimming for a specific piece of information, the child did suggest "just looking" rather than reading. In addition, this reader also realized that titles should make sense. However, this child failed to differentiate between reading for a test and reading for fun.

At the third-grade good reading level, many children showed more sophistication in their responses to questions about monitoring comprehension. For example, one *third-grade good reader* responded as follows.

Q. Knowing when ready to write test.
A. I would know when I was ready because I would know if I knew it good and . . . you know.
Q. How would you know that? Would you just know?
A. I think it would just be like, you know, understanding the things so I would remember it.
Q. Telling parents how test was.
A. Well, because it would be up to you if it was hard or it was easy. Because if you knew it, it would be easy for you. If you didn't, it would be a little hard.
Q. So you could tell, could you, whether it was hard or easy?
A. Yeah.
Q. How?
A. I would tell because if you really studied the test, and after, probably, you'd get perfect in your test because you understood it and answered the questions of the test.
Q. So you would have a pretty good idea before you got the test back?
A. Yeah.
Q. New game. Teaching friend. Could you do it?
A. No.
Q. What would you need to know before you could teach somebody?
A. How I should know how to play.
Q. What sorts of things might you need to ask somebody?
A. Like what do you use, and how you play it, because if you didn't see the game before, then it would be pretty hard.
Q. How could you tell when you knew enough about game?
A. When, uh, like you know how to play it and you know how to do the things in it. So you would probably remember.

In addition, the third-grade good reader has more ideas of different strategies that could be used if comprehension was found faulty. For example, the same third-grade good reader talked about strategics in the following manner.

Q. Reading for a test.
A. Well, I try to keep it in my head and if I forget something, I try to . . . like I read the sentence over and practise it until I know that it's in. Like I read the story and if I still don't quite understand some parts, I go back and read the whole story again.
Q. Read same way for fun?
A. No, because on a test you got to remember it and just for fun, it's . . . you know.
Q. Anything to make what you are reading easier to remember?
A. No.
Q. Story. Name of place where people lived.
A. I would read it and see it, and when it came to the place where I found

the people, where they lived, then I would keep reading and I would try to keep it in my head.

Q. So you would read the whole story then.

A. Yeah.

Q. Do you think that would be the best way?

A. Mm. 'Cause then maybe that's not the real place, and maybe it's somewhere and you don't read it, and it's wrong. That's about it.

Q. Remembering story to tell a friend.

A. I would remember what the things happened in there, and all the characters, so that they would know, so they could tell it to other people.

Q. How much of it would you remember?

A. I don't know. Maybe a little bit. I'm not too good at remembering things.

Q. Thinking of a title.

A. I would think of the title, what it meant, what it said in the story and, uh, the meanings of it in a story so that you could think of it afterwards.

As can be seen, this third-grade good reader has a good grasp of how to monitor comprehension and of what possible strategies can be used in different situations. For example, this reader knew that rehearsal is a necessary part of studying, and if comprehension fails, then the student should go over the material again. Of particular note is his response and rationale for the skim question. This type of caution toward the skim strategy seemed to be typical of good readers. Unfortunately, the *sixth-grade poor reader* often failed to show the same level of sophistication. One example follows.

Q. Knowing when ready to write test.

A. When I know the test, I studied a lot so I knew almost every answer probably.

Q. So you would just know that you knew? OK. Telling parents how test was.

A. Well, if it was hard, I would, uh, be waiting for a long time, and if the test was easy, I'd just put in the answer or something.

Q. So you would know whether or not the questions took you a long time?

A. Yeah. That's it.

Q. New game. Teaching friend. Could you do it?

A. If I knew the game, I would, but if I didn't know the game, I would look at the rules if there were any there.

Q. What would you need to know before you could teach somebody?

A. I wouldn't know anything.

Q. No. What would you need to know?

A. I would need to know how to play the game, if I didn't know it.

Q. Like what sorts of things?

A. Maybe if I learned on something and I didn't understand it, I wouldn't know what to do.

Q. How could you tell when you knew enough about the game?

A. I would know the game off mostly.

In addition, the same sixth-grade poor reader had a limited knowledge of various possible strategies to use when comprehension fails.

Q. Reading for a test.
A. I study it.
Q. Read same way for fun?
A. No.
Q. Anything to make reading easier?
A. I use the ruler.
Q. To follow along?
A. Yeah.
Q. Anything else?
A. I would read the story over a couple of times.
Q. Story. Name of place where people lived.
A. I would read it.
Q. Best way to find the name of the place?
A. No, when I'm reading it along, I'd probably find it and then I'll put my finger on it and put it down, what's the name, so I won't forget it probably.
Q. Any other way?
A. Um, or put a line over it and just continue on with the story.
Q. Remembering story to tell a friend.
A. Do I have to tell it to her? I've got to tell it to her? I would probably memorize it or something.
Q. What would you do to try to memorize it?
A. I would get it into my head.
Q. How much of it would you remember?
A. If it was a long story, I would probably not rememorize half of it.
Q. It would be about half?
A. Yeah.
Q. What if it was a short story?
A. I would know it probably, but I would make a mistake, that's for sure.
Q. So you would remember most of it? Thinking of a title.
A. I would find the best topic for it, like the best thing in the story.

As can be seen, this sixth-grade poor reader knew of few strategies that would aid in comprehension. The concerns expressed seemed to focus not on the use of different reading skills, but with the mechanics of reading (i.e., using a ruler to follow along, writing down answers before forgetting). Perhaps these concerns are indicative of the level of reading at which this child is performing. Fortunately, at the *sixth-grade good reader* level, the child is more apt to exhibit mature knowledge of comprehension-monitoring skills.

Q. Knowing when ready to write test.

A. Um, well, when I . . . if I read the story a few times, and I got the people to ask me questions, and I couldn't answer them all, or had trouble with them, I'd probably wait a while and read some more, and try to remember everything else. And then, once I could, um, remember everything important in it, I'd probably be ready to do the test.
Q. Telling parents how test was.
A. Well, if I didn't get the, what my test was like, my test results, I'd probably, if I thought it was hard, I'd probably tell them that I had difficulty with it, and I'd tell them what the test was like.
Q. How could you tell what it was like?
A. Well, I'd tell them I had trouble with some questions and I might get a low mark from it because I had trouble with them.
Q. How would you know if you had trouble with them?
A. Well, if I had to sit and think about them for a minute, or if I didn't understand them right. That's about it.
Q. New game. Teaching friend. Could you do it?
A. No, I don't think I could, but if my friend and I sort of made up our own rules to it and agreed on them, we could make up our own game from it.
Q. What would you need to know before you could teach somebody?
A. Uh, well, what the game was about and, uh, well, some rules about it and, um, if it had a board that went with it, what certain parts of the board were for, if they had markings on them.
Q. How could you tell when you knew enough about the game to be able to teach somebody else how to play?
A. Well, uh, if I understood the game myself. If I didn't understand parts of it, I wouldn't teach anybody else, I'd wait till I knew how to play it.
Q. How would you know if you understood everything?
A. Uh, well, if I played a game and I knew what to do when it was my turn, if I knew everything about the game, I'd . . .
Q. Then you'd be able to teach.
A. Yeah.

This sixth-grade good reader also was able to explain various strategies that could be used in different situations or if comprehension failed. This same sixth-grade good reader responded as follows.

Q. Reading for a test.
A. Well, first I read it once and then read it over again, and I'll try to remember the important parts, and I'll read it a few times and then I'll get some people to ask me questions on it.
Q. Read same way for fun?
A. No, sometimes I just read it once or twice and then, uh, just look at the pictures in the book.
Q. Anything to make what you are reading easier to remember?

A. Well, uh, you mean if I was going to get a test?
Q. Yes.
A. Well. I could try to underline the important parts. Then I could go over it again, trying to remember the parts I underlined.
Q. Anything else that you might do?
A. No, I don't think I'd do anything else.
Q. Story. Name of place where people lived.
A. Well, I'd first, uh . . . sometimes they have a caption at the top before the story that tells you where it takes place, and about the people, and I'd probably read that first, and if it didn't have it in that, I'd read the whole story.
Q. You'd read the whole story then?
A. Yes.
Q. Anything else that you might do?
A. Well, maybe I might look if there was pictures in it; maybe I could get an idea from the—looking at the people, the way they're dressed and the way the town looks, wherever they live.
Q. Anything else?
A. No.
Q. Remembering story to tell a friend.
A. I think I'd remember all the exciting parts and, um, the main parts, just the main parts in it.
Q. How much of it would you remember?
A. Well, um, probably the beginning and the middle and the end. Not all the, like the little parts in it.
Q. Not the little parts, but the main ideas?
A. Yeah.
Q. OK. Thinking of a title.
A. Well, I'd think, uh, what happened in the story. If it was some kind of adventure story, I'd probably have it sort of an adventurous title. And if it was some kind of, uh, mystery story, I would probably have something that would do with mysteries. It would probably be, maybe, if it's about a person's life, I'd probably have the name of the person. But if it's maybe a name of a car or something, I'd probably have the main thing happened, sort of get an idea from that, and have the title made up from that.

As can be seen, the sixth-grade good reader was quite able to suggest various strategies for reading for different purposes. When faced with a situation, the older/better reader is more apt to put several strategies together into a "plan-of-attack." Less probing is needed from the interviewer. It is curious to note that in spite of this mature knowledge, the older/better reader is often hesitant to use (or suggest) a skim strategy when answering the "name of place" question. In explanation, one sixth-grade good reader suggested the following.

A. Um, I would just, like, just go over it real fast but read all of it, and then as soon as I came to it, then I'd just stick it in my mind and then just read the rest of the story, just scanning.
Q. Best thing to do?
A. Well, yeah, as long as it's not a trick one and you have to have everything.
Q. Know everything in the story instead of just the place?
A. Yeah.

Others, too, suggested that the teacher might ask more questions or that the people in the story might move, so if you didn't read the rest of the story, your answer might be wrong. In fact, many older/better readers suggested that they would like to know what happened for their own interest and would resort to a skim strategy only if pressed for time. However, considering the contrast between the rate of reading of good and poor readers, it is quite possible that good readers are rarely pressed for time in a normal classroom situation.

In summary, it appears that the older/better reader not only knows more about monitoring comprehension, but also knows about different strategies that can be used if comprehension breaks down. On the other hand, the younger/poorer reader shows a definite lack of knowledge about these important skills.

General Discussion of Comprehension and Strategies

We feel that the older/better reader has a greater ability both to perform the component skills of mature reading and to verbalize about the use of these skills. In terms of the development of reading skills, younger/poorer readers may have two problem areas. First, they are *less* likely to change reading strategy to meet the demands of the situation, and second, they are *less* able to assess comprehension and predict accuracy. One possibility is that an inability to monitor comprehension results in a lack of awareness that a change in strategy is needed. Even if younger/poorer readers realized their inaccuracy, it appears that they would have little idea of how to improve comprehension. This possibility was supported by the interview data.

Overall, the younger/poorer reader did less well on the performance, verbal (traditional "meta"), and metacognitive measures of reading skills with which this study was concerned. However, we would urge the reader *not* to conclude that the younger/poorer readers have no competencies, but rather that they have less competence in each area than the older/better readers. For example, all children remembered more important sentences than less important sentences. However, there was a tendency among younger/poorer readers not to recognize problems and/or not to know how to correct them.

In summary, it seems reasonable to conclude the following. (1) Performance on mature reading skills such as comprehension and strategies increases with grade and with reading ability. (2) The ability to monitor comprehension (and predict accuracy) and to make appropriate verbalizations about comprehension and strategies increase with grade and with reading ability. (3) In terms of prediction, performance is a more useful measure than verbalization, both for comprehension and for strategies. (4) Only the older/better reader is a mature metacognizer in the strictest sense, being high on both performance and verbalization skills.

CHAPTER 5

Overview of "Reading Skills" and Preview

As the reader may recall, one of our original aims in doing this study was to examine the overall relationship between various types of reading skills. In order to examine the relationship between each type of reading skill and general reading ability at each grade, we entered the computed scores, along with nonverbal IQ, into a stepwise multiple-regression equation to predict reading ability at each grade. The computed scores that were included in these analyses were the performance measures of decoding, comprehension, and strategies (efficiency), and the verbalization measures of decoding, comprehension, and advanced strategies. The order of entry and the amount of incremental variance accounted for by each score at each grade is presented in Table 5-1.

At the third-grade level, the multiple-regression equation accounted for 69.52% of the variance (R = .8338). As can be seen in Table 5-1, decoding performance was the best predictor of reading performance at the third-grade level, accounting for 54.52% of the variance. Strategies (efficiency) performance and comprehension performance appear to be minimally important predictors. In addition, the reader should note that nonverbal IQ and all of the verbalization measures combined account for less than 1.50% of additional variance.

At the sixth-grade level, the multiple-regression equation accounted for 80.56% of the variance (R = .8975). As can be seen in Table 5-1, strategies (efficiency) performance was the best predictor of reading ability at this level, accounting for 62.90% of the variance. In addition, comprehension performance and decoding performance appear to be minimally important as predictors. The reader also should note that nonverbal IQ and all of the verbalization measures combined account for less than 3.00% of additional variance.

In a predictive sense, it appears that at the third-grade level, decoding performance is the best predictor of reading ability as measured by the Gates-MacGinitie Comprehension Subtest, whereas strategies (efficiency) performance is the best predictor at the sixth-grade level. Since decoding (for third-grade

Table 5-1. Reading Skills Computed Scores: Predicting Reading Ability

	Grade 3		Grade 6	
Variable	Order	Variance	Order	Variance
D-P	1	54.52%	3	4.27%
C-P	3	8.15%	2	10.80%
S-P	2	5.38%	1	62.90%
D-V	6	0.13%	5	0.51%
C-V	4	0.93%	6	0.03%
S-V	5	0.27%	7	0.03%
IQ	7	0.13%	4	2.02%
Total		69.52%		80.56%

children) is the "best" predictor, it would appear that decoding skills play a more significant role than other reading skills in determining third-graders performance on the Gates-MacGinitie Test. A similar statement can be made for strategies (efficiency) at the sixth-grade level. If the Gates-MacGinitie Comprehension Subtest can be inferred to represent "reading ability" in a more general sense, then further speculations can be suggested from the results of these analyses. At the third-grade level, reading ability could be described as the ability to use decoding strategies. However, by the sixth-grade level, reading could be characterized by a level of comprehension that indicates efficient use of reading skills given the length of time spent reading. It would seem that to predict reading ability successfully for each of the two grades, one should use different tests for each grade. When the appropriate measure is used, a large proportion of the variance in reading ability can be accounted for in each grade.

In summary, it appears that the following conclusions can be made. (1) The ability to use the reading skills of decoding, comprehension, and strategies increases with grade and with reading ability. (2) The ability to make mature verbalizations about the reading skills of decoding, comprehension, and strategies increases with grade and with reading ability. (For decoding, the increase with reading ability was not found in all of the items.) (3) In each case, performance measures appear to be more useful than verbalization measures in predicting reading ability. (4) In the strictest sense, taking into account both performance and verbalization, the frequency of metacognizers increases with grade and with reading ability for decoding, comprehension, and strategies. In addition, there is a tendency among sixth-grade poor readers to mimic (a coping strategy). (5) Of all the reading skills measured in this investigation, the best predictor of reading ability is decoding performance for third-graders and strategies (efficiency) performance for sixth-graders.

Since one of the original concerns of this study involved examining how different stages of development affect the reading process, the question that we would like to consider now is the following: Are there developing cognitive processes that may account, at least in part, for the different types of readers found in each grade? It has been argued in an earlier chapter that cognitive processes such as language, attention, and memory affect the development of reading skills. In order to understand more completely the developmental aspects of the reading process, it is important to examine what the child has that he or she brings to the reading situation and what aspects of that child's cognition are still developing. We felt that there are three general areas that should be considered: language, attention, and memory. The child must understand some form of language in order to be able to have some way in which to decode the written symbols. The child also must be able to attend to the task, as well as be able to identify and attend to important elements of the written array. He or she must do both of these in order to understand what has been read. Finally, the child must be able to retain these important elements in order to make any use of the perceived information. Therefore, it seems worthwhile at this point to examine both the cognitive and metacognitive aspects of these component processes as they are developing during middle childhood, and to examine their relationship with the components of reading as described earlier.

CHAPTER 6

Cognitive and Metacognitive Aspects of Developmental Processes

As explained in the introductory chapter, it would be difficult to discuss the acquisition of reading skills without considering the development of basic psychological processes that may affect reading. Rather arbitrarily, we decided to include in our study both cognitive and metacognitive measures of language, memory, and attention (which we will treat primarily in terms of "use" of important information). A brief introduction to what we know about each of these areas follows.

Language

Parents and teachers alike will realize that it would be almost impossible to claim seriously that the development of language abilities has nothing to do with reading. Since reading involves extracting meaning from a written form of language, many researchers and educational consultants have argued that the two skills must relate in some way. In fact, many researchers have studied reading primarily, if not exclusively, from a linguistic point of view (e.g., Goodman, 1969; Smith, 1971). However, it would be presumptuous of us indeed to try to give a capsule view of the issues surrounding language, metalanguage, and their relation to reading in the next few paragraphs. Rather, we would prefer to refer the reader to some of the many excellent sources on the subject (e.g., Dale, 1972; de Villiers & de Villiers, 1978; Kavanagh & Mattingly, 1972; Waterhouse, Fischer & Ryan, 1980).

In spite of a general professional concern for relations between language and reading, the simultaneous examination of both cognitive and metacognitive aspects of language development in relation to reading makes our investigation unique.

Theoretically, the cognitive aspects of language that we feel are important to reading include both semantics and the use of grammatical rules. Metacognitive aspects of language involve the ability to use language skills and to know that there are different ways to say things (flexibility) and that there are rules that make language "sound right." Both aspects of language are important if the reader is to be able to decode not only to sound, but also to meaning.

Important cognitive aspects of language can be measured by vocabulary, the ability to identify concepts used in reading (such as word, sentence), and the ability to use grammatical rules to make appropriate sentences. An index of metacognitive development can be provided, for example, by an explanation of the changes made to anomolous sentences in order to make them read correctly.

With these points and distinctions in mind, the children were given a series of language performance tasks (in an attempt to assess cognitive aspects of language development) and were interviewed in an attempt to assess their knowledge about these skills.

Memory

We suspect that we will have little difficulty convincing our readers that memory is another factor that might affect reading and influence the development of reading skills. Obviously, a reader must be able to remember what has been read in order to make use of the material at a later time. In fact, many psychologists (e.g., Mackworth, 1972) have developed models of reading that give a very major role to memory. For excellent reviews of the relevant literature on memory, metamemory, and the relation of these to reading, see, for example, Baker and Brown, in press; Flavell, 1977; Gibson and Levin, 1975; and Singer and Ruddell, 1976. Once again, the simultaneous measures of memory, metamemory, and reading make the following data of considerable interest.

Theoretically, we feel that the cognitive aspect of memory that is important to reading is the ability to retain information for later use. The reader must know that different memory strategies might be more efficient given different specific purposes. Therefore, we logically conclude that aspects of memory and metamemory are important for comprehension. Memory, in a cognitive sense, can be measured by performance on recall or recognition tasks. In a metacognitive sense, memory can be measured by the use of mnemonic strategies plus the ability to reflect on the efficiency of those strategies given a specific situation.

Use of Important Information: Attention

If a child is reading a passage for a particular purpose, especially if that purpose is a sophisticated reading strategy such as studying, then the reader must be able to direct attention to important information (e.g., Miller, in press). Our readers will note that we are not using the term "attention" here in the traditional perceptual sense (e.g., Singer & Ruddell, 1976). Rather, it is being used to explain the ways in which the child uses his or her knowledge of important information. A reader who does not "pay attention" to the relevant cues certainly will have difficulty decoding symbols to sounds. Similarly, a reader who does not "pay attention" to the words will have difficulty comprehending what has been read.

The importance of attentional capacities to reading has been treated by a number of writers. For example, LaBerge and Samuels (1974) suggested that attention is a key factor in reading, and they argued that it is only when the skills involved in reading become fully "automatized" that a person can read fluently. In addition, Willows (1974) suggested that poor readers are more easily distracted from the reading task than good readers. We would like to argue, however, that the role of attention in mature readers may take on a more complex significance. For example, Brown and Smiley (1977) found that younger children (ages 8 and 10 years) were less able than older children (12 and 18 years) to identify important units of text. Being able to identify important units of text may be an important prerequisite for directing attention appropriately when using advanced reading strategies.

In this investigation, we admit that a fairly nontraditional position has been taken with regard to attention. In a cognitive sense, the view of attention that we adopted was that in any given situation, certain cues, or information, could make the task easier if attention was directed appropriately (e.g., use of redundant information in a *search* task). In addition, knowledge about relevant information, the way in which certain information aids in any given situation, and the importance of various pieces of information were seen as important metacognitive aspects of attention. Cognitive aspects of attention can be measured, for example, by matching tasks, sorting tasks, and search tasks, whereas metacognitive aspects can be measured by the use of attention skills plus the ability to reflect on attention skills (e.g., what information is important in any given situation, where should attention be directed, etc).

Method, Results, and Discussion

Performance Items

In this section we present the results of our investigation of the cognitive and metacognitive aspects of these processes (language, memory, and attention) and examine their relations to reading. Summaries of the statistical analyses of all of the developmental items are presented in Appendix N.

Language

The items used to measure cognitive aspects of language, the child's actual language ability, are summarized in Table 6-1 (and presented in full in Appendix O.) The first item, vocabulary, was a measure of the child's receptive compre-

Table 6-1. Summary of Language Performance Items

Task L-P-1:	Vocabulary.
Task L-P-2:	Production of letter.
Task L-P-3:	Production of word.
Task L-P-4:	Production of sentence.
Task L-P-5:	Recognition of letter.
Task L-P-6:	Recognition of word.
Task L-P-7:	Recognition of word.
Task L-P-8:	Recognition of sentence.
Task L-P-9:	Errors in sentence recognition.
Task L-P-10:	Recognition of word/nonword.
Task L-P-11:	Recognition/correction of grammatical errors.
Task L-P-12:	Active/passive transformations.
Task L-P-13:	Production of homonyms.
Task L-P-14:	Latency of word/nonword judgment.
Task L-P-15:	Latency of word/nonword judgment.
Task L-P-16:	Latency of word/nonword judgment.

hension of written words. Items 2 through 10 and 14 through 16 were used to assess the child's ability to use language concepts important to reading. Items 12 and 13 were used to assess the child's flexible use of language. Means for the individual language performance items are presented in Table 6-2.

Two items were deleted from further analyses. These were production of a word and production of a letter. All children successfully produced a word, and only three children, all third-grade poor readers, made errors on the letter-production task, indicating a ceiling effect. (Two of the children produced a number and one left a blank.)

The remaining items were designed to assess the child's vocabulary, ability to recognize and produce language concepts, and ability to recognize and correct grammatical errors. The items, plus nonverbal IQ, were analyzed in a grade by reading ability analysis of variance. The multivariate analysis of variance showed changes with grade and with reading ability. In addition, the grade by reading ability interaction was significant. All of the univariate F's with their probability levels are presented in Appendix N.

Vocabulary scores, production of a sentence, recognition of letters and words (Downing & Oliver's, 1974, stimuli), word/nonsense words, grammatical errors and corrections, and the ability to make active/passive transformations and to produce different meanings for homonyns all increased significantly with grade. The number of errors in the sentence-recognition task decreased significantly with grade, as did the response latency of word/nonword judgment for "tdet" and "meff."

In addition, the vocabulary scores, grammatical acceptability and corrections, the ability to make active/passive transformations and to produce different mean-

Table 6-2. Language Performance Items: Means

Measure	Maximum Score	Grade 3 Poor	Grade 3 Average	Grade 3 Good	Grade 6 Poor	Grade 6 Average	Grade 6 Good
L-P-1	—	42.13	51.92	58.79	45.50	54.04	60.88
L-P-2	1	0.88	1.00	1.00	1.00	1.00	1.00
L-P-3	1	1.00	1.00	1.00	1.00	1.00	1.00
L-P-4	3	2.00	2.33	2.50	2.54	2.79	2.83
L-P-5	7	5.92	6.63	6.58	6.96	7.00	6.92
L-P-6	7	6.46	6.92	6.96	7.00	7.00	7.00
L-P-7	10	8.08	9.08	9.46	8.88	9.63	9.71
L-P-8	5	3.33	3.54	3.75	3.63	3.75	3.92
L-P-9	5	2.71	2.17	2.08	1.25	1.04	0.21
L-P-10	3	2.08	2.29	2.54	2.33	2.79	2.75
L-P-11	16	9.25	10.42	12.88	13.00	14.13	15.00
L-P-12	6	2.63	3.58	3.83	4.63	4.67	5.83
L-P-13	10	6.25	7.63	8.50	8.50	9.17	9.75
L-P-14	—	0.61	0.56	0.44	0.29	0.38	0.32
L-P-15	—	0.51	0.48	0.39	0.43	0.32	0.40
L-P-16	—	0.16	0.10	0.01	0.01	0.01	0.01

ings for homonyms all increased with reading ability, differentiating between good, average, and poor readers. Production of a sentence, recognition of words (Downing & Oliver's, 1974, stimuli), and judgment of word/nonword increased with reading ability, differentiating poor readers from average and good readers. Also, the number of errors in the sentence-recognition task decreased with reading ability, differentiating good readers from poor readers. Measures of recognition of a word, recognition of a sentence, and the response latency of word/nonword judgments for "meff" showed no difference.

We also assessed the relationship between each performance item and reading ability at each grade level by means of correlations and partial correlations, controlling for nonverbal IQ. The results (shown in Appendix N, Table N-2) indicated that the language performance items and reading ability are correlated, and the relations remain when the variance attributed to nonverbal IQ is removed. Therefore, it can not be argued that the relationship between reading ability and the language performance items is caused solely by nonverbal IQ, at least as measured in this investigation.

We also computed for each child a single score to represent that child's overall level of achievement on language performance. For the performance computed score, all language performance items were included except production of a letter (L-P-2) and production of a word (L-P-3), both of which had been deleted earlier. The errors in the recognition of sentences task (L-P-9) and the latency of response measures for "tdet," "meff," and "stone" (L-P-14, L-P-15, L-P-16) also were deleted since these measures were expected to decrease with grade and with

Table 6-3. Summary of Memory Performance Items*

Task M-P-1:	Recall of list of words, no delay.
Task M-P-2:	Recall of list of words, 1-minute delay.
Task M-P-3:	Recall of list of words, 1-minute inference task.
Task M-P-4:	Recall of list of words, category cluster, no delay.
Task M-P-5:	Recall of telephone number.
Task M-P-6:	Recall of pictures of items from story.
Task M-P-7:	Recall of clusterable pictures.
Task M-P-8:	Recall of names.

*Items 5-8 were embedded in the interview part of the study.

reading ability. (The computed score was calculated to represent the increase in ability with older children and greater reading skill. If variables that were expected to decrease with these two variables were included, the older/better reader would have been penalized unless the data were transformed. We elected not to do so.) The computed score of language performance, presented in Table 7-1 in a later chapter, increased with grade and with reading ability (Appendix N). More will be said about these scores later.

The results of the analyses on the performance items reaffirmed what many researchers have suggested: performance on language skills increases with grade and with reading ability. However, it is worth noting that even younger/poorer readers have a certain level of language competence. They can produce a letter and a word; they can recognize a word in arrays of letters, numbers, shapes, and two-word clusters, and they can recognize sentences. As the items became slightly more difficult, the poor readers were differentiated from the average and good readers by production of a sentence, recognition of words using Downing and Oliver's stimuli, and judgments of word/nonword. It was only with the more difficult items that the average and good readers were differentiated. Vocabulary, recognition and correction of sentence errors, active/passive transformations, and production of meanings for homonyms differentiated all three reading levels. Overall, the results of the performance items confirmed what was expected on the basis of previous findings (Downing & Oliver, 1974; Ehri, 1976; Ryan, McNamara & Kenny, in press).

Memory

The items used to measure cognitive aspects of memory are presented in brief in Table 6-3 (and in full in Appendix P). For items 1 to 4, the children were asked to try to remember a list of words and then reproduce them. In each case, the study period was two minutes. The different conditions are described briefly below. In each case, a score of one point was given for each word correctly recalled, with a total possible of 12. The words used in these items were chosen from the Lorge-Thorndike frequency list. All had an AA listing. Items 1 through 4 were designed to assess the child's ability to intentionally remember visually

Table 6-4. Memory Performance Items: Means

Measure	Maximum Score	Grade 3			Grade 6		
		Poor	Average	Good	Poor	Average	Good
M-P-1	12	4.79	7.17	7.29	10.00	10.25	11.13
M-P-2	12	5.04	6.54	7.63	8.79	9.83	10.75
M-P-3	12	3.54	5.50	5.42	7.83	9.13	10.00
M-P-4	12	6.71	8.54	9.67	11.08	11.42	11.92
M-P-5	7	0.83	1.00	1.76	2.13	2.83	2.91
M-P-6	7	6.25	6.42	6.17	6.42	6.88	6.83
M-P-7	9	4.75	5.38	5.29	6.08	6.83	7.50
M-P-8	8	1.82	1.91	2.13	2.10	2.55	2.96

presented verbal material. Items 5 through 8 were designed to assess the child's ability to remember incidental information. Means for all the individual items are presented in Table 6-4.

All performance item (except for two) and nonverbal IQ were analyzed in a grade by reading ability multivariate analysis of variance. The two items that were deleted were recall of telephone number and recall of names of children participating, both of which were incidental information from the interview. These items were dropped because of missing data where the interviewer failed to ask for the information. The multivariate analysis of variance showed a change with grade and a change with reading ability (Appendix N, Table N-3). As can be seen in Table N-3, all of the memory items that were entered into the analysis indicated that recall (both intentional and incidental) increased with grade. In addition, all but one item showed that recall increased with reading ability. The only item that failed to show a main effect of reading ability was the incidental recall of pictures of items in a story. The other incidental recall task with clusterable pictures showed a significant difference between poor and good readers. The remaining tasks indicated that the poor readers were differentiated from both average and good readers. The results lead us to conclude that if memory performance and reading are related, it is mainly a problem for poor readers. The only significant interaction between grade and reading ability involved the immediate recall of a list of words. At the third-grade level, poor readers recalled words less frequently than average and good readers. However, there were no differences on this task at the sixth-grade level.

We also assessed the relationship between each performance item and reading ability at each grade level by means of correlations and partial correlations, controlling for nonverbal IQ (Appendix N, Table N-4). Memory performance is related to reading ability, and this relationship is maintained when the variance attributed to nonverbal IQ is removed. This is especially true at the lower grade level.

For the computed scores, all memory performance items were combined except

Table 6-5. Summary of Attention Performance Tasks

Task A-P-1:	Matching, letters.
Task A-P-2:	Matching, one word.
Task A-P-3:	Matching, sentence.
Task A-P-4:	Use of cues for sorting, conceptual categories.
Task A-P-5:	Use of cues for sorting, rhyming categories.
Task A-P-6:	Recognition of important units.
Task A-P-7:	Identification of less important sentences as important.
Task A-P-8:	Production of important units (telegraphic note).
Task A-P-9 and Task A-P-11:	Redundancy-search tasks.
Task A-P-10 and Task A-P-12:	Search patterns.

recall of telephone number (M-P-5) and recall of names of children at the party (M-P-8), both of which had been deleted. The computed scores on memory performance, presented in Table 7-1 in a later chapter, increased with grade and with reading ability. More will be said about these scores later.

The results of the analyses of the performance items reaffirmed what researchers have suggested, that performance on memory tasks increases with grade and with reading ability. However, we must note that the younger/poorer reader does have a certain level of memory competence. In the intentional recall tasks, it was very rare that the children could recall none of the words. Also, none of the tasks appeared to be difficult enough to sharply differentiate one reading group from another.

Use of Important Information: Attention

The items summarized in Table 6-5 (and presented in full in Appendix Q) were used to measure cognitive aspects of attention. Items 1 through 5 were designed to assess the child's ability to use information, or cues, to successfully match or sort different units. Items 6 through 8 were designed to assess the child's ability to recognize the importance of various units. Items 9 through 12 were designed to assess the child's ability to use information to aid in a search situation. Means for all the individual items are presented in Table 6-6.

All the items, plus nonverbal IQ, were analyzed in a grade by reading ability multivariate analysis of variance. The multivariate analysis of variance showed a change with grade and a change with reading ability. In addition, the overall grade by reading ability interaction was significant. Performance on all of the attention items improved with grade except for the identification of unimportant units as important (A-P-7), which decreased with grade, as expected. In addition, some of the univariate analyses showed a main effect of reading ability. The identification of important units increased with reading ability and differentiated all three reading groups. Matching sentences and production of important units

Table 6-6. Attention Performance Items: Means

Measure	Maximum Score	Grade 3			Grade 6		
		Poor	Average	Good	Poor	Average	Good
A-P-1	7	6.33	6.46	6.63	6.63	6.92	7.00
A-P-2	7	6.50	6.92	6.75	6.88	7.00	7.00
A-P-3	5	3.00	3.50	4.25	4.50	4.67	4.79
A-P-4	12	7.25	9.88	10.67	11.71	11.42	12.00
A-P-5	12	5.42	7.79	7.00	9.96	9.92	11.71
A-P-6	3	1.71	2.17	2.75	2.75	2.96	3.00
A-P-7	8	2.29	2.13	1.00	1.38	0.96	1.13
A-P-8	5	0.79	1.17	2.25	3.42	3.71	4.58
A-P-9	18	14.63	13.08	14.71	16.75	16.29	16.17
A-P-10	2	0.67	1.08	1.33	1.83	1.58	1.88
A-P-11	18	11.58	12.17	11.50	14.79	15.29	15.92
A-P-12	2	1.21	1.38	1.63	1.96	1.92	2.00

both increased with reading ability, differentiating good readers from average and poor readers. The ability to use cues in conceptual clustering also increased with reading ability, differentiating poor readers from average and good readers. Use of a systematic search on the redundancy task and use of cues in rhyming clustering both increased with reading ability, differentiating poor readers from good readers. Finally, the tendency to identify unimportant sentences as important units decreased with reading ability, differentiating poor readers from good readers.

Several of the items presented different patterns of results for reading groups in each grade (i.e., search pattern on redundant-search task [A-P-9], use of cues in conceptual clustering [A-P-4], and identification of important units [A-P-6]). Generally, the distributions appeared much flatter for the sixth grade than for the third grade. The results of the search pattern on the redundancy-search task (A-P-9) indicated no difference at the sixth-grade level, but at the third-grade level poor readers used a systematic search less often than average and good readers. The identification of important units (A-P-6) showed no difference between readers at the sixth-grade level, but at the third-grade level, poor readers identified fewer important sentences than average readers, and average readers identified fewer important sentences than good readers. Again, in the conceptual clustering task (A-P-4), the results indicated that there was no difference at the sixth-grade level, but at the third-grade level poor readers did not cluster as well as average or good readers.

We also assessed the relationship between each performance item and reading ability at each grade level by means of correlations and partial correlations, controlling for nonverbal IQ (Appendix N, Table N-6). Some items (Table N-6) have a fairly strong relationship with reading ability, and this relationship is minimally affected by the removal of variance attributed to nonverbal IQ.

For the attention performance computed score, all performance items were included except for the identification of less important sentences as important. This item was deleted because it decreased with grade, as expected. The computed score for attention performance, presented in Table 7-1 in a later chapter, increased with grade and with reading ability. In addition, grade interacted with reading ability. At the third-grade level, poor readers had a lower level of achievement than the average readers, and average readers had a lower level of achievement than the good readers. At the sixth-grade level, the difference was much smaller, but poor readers achieved a lower level on attention performance than the good readers. More will bc said about these scores later.

The results of the analyses of the attention performance items reaffirm that attention skills do develop with age. It was the case that each item showed a main effect of grade, and most of the differences were highly significant. While the third-grade children were having some difficulty with some of the attention tasks, the sixth-grade children were reaching a ceiling on many of the items.

In addition, there were various items that differentiated the reading levels. The most impressive items in this regard were the items that asked the child to identify and produce important units within a story. This result substantiates what Brown and Smiley (1977) have suggested, that the ability to identify important units increases with age and is important to advanced comprehension skills.

Other attention skills also seem to be moderately important, but it could be that some of these skills (such as use of cues in clustering, matching, and searching) are ones usually associated with primary reading, and so most of these children have acquired them at least to a certain degree.

Verbalization Items

Language Verbalization Items

The following interview items were designed to assess each child's knowledge of language concepts used in reading, language flexibility, and grammatical acceptability. The number of children in each grade and reading level giving any particular level of response for each of 21 interview items is shown in Table 6-7 with the appropriate overall chi-square value and probability level. The items and scoring criteria are presented in full in Appendix R.

Question L-V-1: What can you tell me about words?

Most children, even third-grade poor readers, could give at least one characteristic of a word, while sixth-grade good readers were able to give a combination of characteristics. The ability to produce a sophisticated response to this item increased with grade and with reading ability. The effect of reading ability was evident mainly in an increase between average and good readers.

Question L-V-2: Where do we use words?

Third-grade poor readers also were able to verbalize the fact that words are used in at least one mode (speech or writing), while by the sixth-grade reading level, both modes usually were mentioned. The ability to produce a sophisticated response to this item increased with grade.

Question L-V-3: What can you tell me about sentences?

Similar to how they responded to the equivalent question about words, third-grade poor readers could give at least one characteristic of sentences, while sixth-grade good readers were able to think of a combination of characteristics. The ability to give a sophisticated response increased with grade and with reading ability. An important increase appeared between sixth-grade average and good readers.

Question L-V-4: Where do we use sentences?

Most children realized that sentences could be used in any written form. However, there was an increase in the frequency of sophisticated responses with grade.

The following 15 questions were incorporated under the same introductory cover story. The first three were designed to assess the child's knowledge about words. All verbal stimuli was presented to the child in written form (typed on three-by-five cards) and repeated verbally by the experimenter.

Question L-V-5: If Johnny wrote "tdet" and showed it to you, would you mark it right or wrong? What makes you think it is/is not a word?

Question L-V-6: If he wrote "meff," would you mark it right or wrong? What makes you thing it is/is not a word?

Question L-V-7: If he wrote "stone," would you mark it right or wrong? What makes you think it is/is not a word?

Almost all of the children knew that "stone" was a word and could give an adequate answer as to why this was so. However, approximately half of the children at the third-grade poor and average levels and at the sixth-grade poor level claimed that "tdet" and/or "meff" was/were correct. The ability to give a sophisticated response increased with grade for "tdet" and with reading ability for "meff." This increase in reading ability was evident particularly between average and good readers.

Question L-V-8: Suppose that you were not sure whether to mark one of Johnny's words right or wrong. Is there anything you could do so that you would be sure?

If faced with a situation where they didn't know whether or not a group of letters was or was not a word, third-grade poor readers could think of nothing that they could do so that they would be sure. Most other children could think of at least one thing that they could do (ask teacher, check dictionary), and the

Table 6-7. Language Verbalization Items: Frequencies of Scores

		L-V-1	L-V-2	L-V-3	L-V-4	L-V-5	L-V-6	L-V-7
Grade 3								
Poor	0	7	3	7	7	8	13	1
	1	17	19	17	3	14	10	13
	2	0	1	0	13	2	1	10
Average	0	7	4	5	7	5	12	0
	1	17	18	16	2	16	11	13
	2	0	2	3	15	3	1	11
Good	0	6	0	3	4	4	6	0
	1	17	20	19	3	13	14	9
	2	1	4	2	17	7	4	15
Grade 6								
Poor	0	5	3	4	1	1	13	0
	1	17	17	16	4	9	11	12
	2	2	4	4	19	14	0	12
Average	0	3	0	1	0	0	4	0
	1	19	14	18	1	11	17	6
	2	2	10	5	23	13	3	18
Good	0	1	0	0	0	2	4	0
	1	15	17	11	2	7	12	9
	2	8	7	13	22	15	8	15
Chi square		27.13**	23.46**	34.96****	23.81**	36.19****	29.26***	12.18

		L-V-8	L-V-9	L-V-10	L-V-11	L-V-12	L-V-13	L-V-14
Grade 3								
Poor	0	14	2	1	0	14	23	8
	1	8	4	19	9	1	1	1
	2	0	6	4	15	9	0	15
Average	0	2	1	0	0	10	20	6
	1	17	7	14	3	2	0	1
	2	2	3	10	21	12	4	17
Good	0	3	0	0	0	4	15	3
	1	15	3	14	6	2	2	0
	2	1	3	10	18	18	7	21
Grade 6								
Poor	0	0	2	0	0	4	11	4
	1	15	6	17	1	3	1	1
	2	8	5	7	23	17	12	19
Average	0	0	0	0	0	6	5	4
	1	17	1	13	3	0	0	2
	2	6	4	11	21	18	19	18
Good	0	1	0	1	0	4	2	4
	1	13	3	8	1	0	0	1
	2	9	1	15	23	20	22	19
Chi square		62.57****	7.83	16.32	15.16**	23.49**	66.76****	6.63

Table 6-7 (*continued*)

		L-V-15	L-V-16	L-V-17	L-V-18	L-V-19	L-V-20	L-V-21
Grade 3								
Poor	0	14	3	11	2	2	13	7
	1	7	6	9	4	3	10	15
	2	3	15	4	18	19	1	2
Average	0	15	0	8	3	1	7	5
	1	1	4	12	3	2	13	8
	2	8	20	4	18	21	4	11
Good	0	5	0	5	2	2	5	6
	1	1	6	14	4	2	17	6
	2	18	18	5	18	20	2	12
Grade 6								
Poor	0	4	0	6	1	0	2	1
	1	4	7	14	4	5	13	9
	2	16	17	4	19	19	9	14
Average	0	3	0	3	2	1	1	1
	1	0	6	16	5	6	16	4
	2	21	18	5	17	17	7	19
Good	0	2	0	3	1	0	0	0
	1	0	3	14	1	1	11	3
	2	22	21	7	22	23	13	21
Chi square		60.79****	18.38*	11.60	5.08	10.99	45.28****	43.29****

Note: Where cell frequencies do not sum to 24, there were small bits of missing data for individual children on specific items. These children were not included in the frequency analyses on those items.

$^{*}p < .05$ $^{**}p < .01$ $^{***}p < .001$ $^{****}p < .0001$

ability to think of two ways increased with grade. In addition, third-grade good readers responded at a significantly lower level than sixth-grade poor readers.

If more than just "stone" was marked right, the child was asked the following question.

Question L-V-9: Which one is Johnny's best word? Why? What makes it the best word?

Of those children failing to differentiate nonsense words (tdet, meff) from real words, most knew that "stone" was the best word, and approximately half could give a reason why this was so.

The following ten questions were presented under the same cover story that was given for the previous questions. The items were designed to assess the children's knowledge of the concept of sentence and grammatical acceptability. Therefore, a question was asked after each sentence to determine whether or not the child was aware of the problem in the sentence and if so, if he could correct it.

Question L-V-10 to L-V-19: Then you asked Johnny to write one sentence. He wrote:

(10) John park to went.
(11) Jane played with her friends.
(12) After school, Bill wented home.
(13) Before Mary could enter the contest.
(14) My favorite dessert is radios with cream.
(15) My favorite TV program are Gunsmoke.
(16) My favorite toothpaste is Crest.
(17) I paid the money by the man.
(18) I gave the cash to the girl.
(19) My favorite breakfast is eggs with bacon.

After each sentence, the child was asked the following questions: Would you mark it right or wrong? What makes it a sentence/not a sentence? (Items 14 through 19 are from M. Dennis, unpublished.)

There was a tendency for everyone to do well on judging the grammatical acceptability of those sentences that were very easy. Everyone knew that "John park to went" was wrong and most could explain why. "My favorite dessert is radios with cream" also was recognized as nonsensical. In addition, most knew that "I gave the cash to the girl" was correct because it made sense. "My favorite breakfast is eggs with bacon" and "My favorite toothpaste is Crest" also were recognized as correct. Conversely, it appeared that one item ("I paid the money by the man") was too difficult, even for the good readers, and consequently showed no systematic effects. Most children recognized that it was wrong, but few could explain that the "by" should have been "to." Those who said the sentence was correct (scored 0) tended to explain the sentence as "I paid the money while standing beside the man."

In judging the grammatical acceptability of sentences, many younger/poorer readers had difficulty identifying a mistake in "After school, Bill wented home." It was only older/better readers who could explain that 'went' did not need an 'ed' like some verbs did. In addition, no one claimed that "Jane played with her friends" was incorrect, but only older/better readers could explain that the sentence made sense. Older/better readers also were better able to recognize errors and justify their judgments in "Before Mary could enter the contest" and "My favorite TV program are Gunsmoke." The results of the chi-square analyses for sophisticated responses are shown in Appendix N, Table N-8.

The following item was designed to assess the children's knowledge of the flexibility of language. An attempt was made to have the children recognize and make changes in the structure of language before asking them about ways to make their language sound different.

Question L-V-20: Can you always say the same thing in different ways? How do you make the same idea sound different?

The ability to give a sophisticated response explaining how one can make ideas sound different increased with grade.

Question L-V-21: How come you can give me two meanings for each of the words we have talked about so far? If I kept giving you words, do you think that you could always give me two different meanings for every word? Why do you think you could/could not?

The ability to explain why one could not always think of two different meanings for every word increased with grade and with reading ability. In addition, poor readers responded at a significantly lower level than average readers, particularly at the third-grade level.

We also assessed the relationship between each verbalization item and reading ability at both grade levels by means of correlations and partial correlations, controlling for nonverbal IQ (Appendix N, Table N-9). Most items are affected only minimally by the removal of variance attributed to nonverbal IQ, indicating that the relationship is not caused solely by nonverbal IQ.

In computing the language verbalization computed score, the item that was deleted from the computation was one that was asked only of those children who had indicated that they felt that "tdet" or "meff" was a word (L-V-9). The language verbalization computed score, presented in Table 7-1 in a later chapter, increased with grade and with reading ability.

Overall, the younger/poorer readers seemed less able to express knowledge about words and sentences than older/better readers. Some could think of very little to say about words and sentences other than that we use them to read and write. In contrast, one sixth-grade good reader explained words and sentences as follows: "Well, all words have a vowel and most of them have consonants, um, and there are long vowels and short vowels in sentences and words, and when you have an "e" at the end, it is usually, like in the word "cake," it makes the "a" long. Sentences have a subject and a predicate and they are a complete thought. At the beginning of them, they always have a capital, and at the end, a punctuation mark."

The ability to produce appropriate justification for judgments of grammatical acceptability also increased with grade and with reading ability. Younger/poorer readers often could not identify mistakes, or if the mistake was recognized, they gave little indication that they knew how to correct it. In addition, the younger/poorer reader showed less knowledge about the flexibility of language and less knowledge about different types of words (in this case, homonyms).

As expected, the older/better reader was more able to make verbalizations about his or her language skills. The older/better reader appeared to know more about words, sentences, and grammatical rules and to exhibit more knowledge about the flexibile use of words. Once again, it is not appropriate to assume that the younger/poorer readers have no knowledge about their language skills. For example, most children were able to recognize correct sentences as correct

Table 6-8. Memory Verbalization Items: Frequencies of Scores

		M-V-1	M-V-2	M-V-3	M-V-4	M-V-5	M-V-6	M-V-7	M-V-8
Grade 3									
Poor	0	10	0	11	3	18	4	7	7
	1	1	20	6	20	5	17	17	3
	2	13	4	7	1	1	3	0	14
Average	0	10	0	8	3	19	4	5	7
	1	0	16	4	18	3	20	18	1
	2	14	8	12	3	0	0	1	16
Good	0	12	0	5	0	16	3	3	7
	1	0	20	6	22	4	21	18	1
	2	12	4	13	2	2	0	3	16
Grade 6									
Poor	0	9	1	3	1	18	2	1	2
	1	1	12	2	19	5	20	19	1
	2	14	11	18	4	0	2	4	21
Average	0	8	0	3	1	12	1	0	4
	1	1	11	0	15	8	18	16	0
	2	15	13	21	8	1	5	8	20
Good	0	7	0	3	0	10	0	1	3
	1	3	9	0	17	10	15	10	0
	2	14	15	21	7	4	9	13	21
Chi square		8.03	24.25*	32.84*	17.91	18.14	25.67*	40.85**	13.76

Note: Where cell frequencies do not sum to 24, there were small bits of missing data for individual children on specific items. These children were not included in the frequency analyses on those items.

*$p < .01$ **$p < .0001$

because they make sense. In addition, they could often recognize an incorrect sentence as wrong, but could offer no explanation for their judgment. It were only the olders/better reader who was able to express a relatively high level of knowledge about their language skills (e.g., they could recognize an incorrect sentence as wrong and could identify and explain the actual mistake).

Memory Verbalization Items

The memory interview items were selected and adapted from the classic Kreutzer, Leonard, and Flavell (1975) monograph on metamemory. The number of children in each grade and reading level giving any particular level of response for each item is shown in Table 6-8 along with the appropriate chi-square value and probability level. The complete list of items and scoring criteria are reported in full in Appendix S.

Question M-V-1: Suppose you wanted to phone your friend and someone told you the phone number. Would it make any difference if you called right away after you heard the number or if you got a drink of water first? Why?

Most children knew that they should call right after hearing a telephone number rather than get a drink of water first. However, not all children could explain why this was so.

Question M-V-2: What do you do when you want to remember a telephone number?

The younger/poorer readers usually stated that they would remember a telephone number by writing it down or by repetition. In contrast, the olders/better reader tended to suggest both strategies or a clustering strategy. The ability to give a sophisticated response to this item increased with grade.

The following item was designed to assess the child's knowledge of the demands of the data on the ability to recall. In this particular case, the knowledge that was being assessed was whether or not the child was aware that hearing a story about pictures would provide a structure that would make the pictures easier to remember.

Question M-V-3: Do you think the story made it easier or harder for the girl to remember the pictures? Why?

The younger/poorer reader tended to suggest that hearing a story would make pictures of items in that story harder to remember. However, by the sixth-grade average reading level, most children were suggesting that the story would make the pictures easier to remember because you could use the theme to aid in retaining the information. The ability to make a sophisticated response to this item increased with grade.

Question M-V-4: What would you do to learn these [clusterable] pictures?

When asked how they would remember clusterable pictures, the younger/poorer readers often suggested looking at them a lot or using a repetition strategy. The tendency to suggest clustering increased with age and with reading ability, and the ability to produce a sophisticated response to this item increased with grade.

Question M-V-5: Why would you do it that way? (Referring to above item.) Is there anything else that you could do?

Most younger/poorer readers could not suggest a second strategy for remembering the same pictures, whereas the older/better reader usually could think of alternatives. The ability to produce a sophisticated response to this item increased with reading ability.

Question M-V-6: Suppose you lost your jacket while you were at school. How would you go about finding it?

When explaining how to find a lost jacket, the younger/poorer reader could suggest possible cues, but it was usually the older/better reader who suggested a systematic search (e.g., go back along path until you find it, remember where you last had it, etc.). The ability to produce a sophisticated response to this item increased with grade.

Question M-V-7: Anything else that you could do? (Referring to the above item.)

The older/better reader could think of more alternative ways to find a lost jacket. The ability to produce a sophisticated response to this item increased with grade and with reading ability.

Question M-V-8: Which friend do you think remembered the most names, the one who went home after the party or the one who went to practice in a play where he met some more children? Why?

Most children knew that interference (hearing more names) would make things harder to remember. The ability to produce a sophisticated response to this item increased with grade.

The relationship between each verbalization item and reading ability at each grade level also was assessed by means of correlations and partial correlations, controlling for nonverbal IQ (Appendix N, Table N-11). There is a relationship between some of the memory items and reading ability at each grade, and the relationship is not affected dramatically by the removal of the variance attributed to nonverbal IQ. In addition, memory items appear to have a stronger affect at the later grade.

The computed score representing the child's overall level of achievement on memory verbalization was based on all verbalization items described above. This computed score on memory verbalization, presented in Table 7-1 in a later chapter, increased with grade and with reading ability.

Overall, the young and, to a certain extent, the poor reader had little idea of the procedures that might help in memory tasks and often could provide no rationale for using memory strategies. However, the older/better reader was more able to give adequate answers to memory problems. This type of reader seemed to be able to identify various possible strategies and could choose among them to optimize efficiency. For example, when asked how he would remember clusterable pictures, one sixth-grade good reader responded as follows.

A. Do I have to remember both yellow corn or just corn?
Q. Just corn. What would you do to learn these pictures?
A. Well, put these together, clothing that could be worn, and the animals.
Q. OK. Why would you do it that way?
A. Well, then they'd be in categories.
Q. That would be easier?
A. Mm-hm.
Q. Anything else?

A. Well, I could put 'em under letters. Like corn, carrots, coat, and chick. That wouldn't be as good as the other way, squirrel and shoes. Mittens, frog, and grapes would be all separate.
Q. Anything else?
A. Well, they could be put in order according to colors: yellow, purple, brown, orange.
Q. But clothing and food and animals would be best?
A. Mm-hm.

As expected, the older/better reader was able to make appropriate verbalizations about his memory processes. The older/better reader seemed to be able to identify the possible strategies and could choose among them to optimize efficiency. These results substantiate those found by Kreutzer, et al. (1975).

Use of Important Information: Attention Verbalization Items

The following item was designed to assess the children's knowledge of information that might be useful in the solution of tasks such as search plans. The number of children in each grade and reading level giving any particular level of response for each verbalization item is shown in Table 6-9 with the appropriate chi-square value and probability level. The complete items and scoring criteria are presented in Appendix T.

Question A-V-1: Suppose you had a lot of books to choose from, like in a library. How would you decide which one you wanted to read [i.e., recognition of important information for a search]?

Most younger/poorer readers could identify cues that they used when picking out books from the library, but it was only the older/better reader who presented these cues in such a way as to suggest a systematic search strategy. The ability to produce a sophisticated response for this item increased both with grade and with reading ability.

The following item was designed to assess the child's knowledge of important cues in evaluating a situation, and his or her knowledge of what information was more important.

Question A-V-2: What would be the best way to decide who was a good reader?

Most children could explain one way of identifying a good reader: having each child read out loud or by giving a reading comprehension test. Having each child read out loud was the most common answer, and giving a combination of ways was rare.

Question A-V-3: Why would that be the best way?

Most younger/poorer readers could give one reason for their method of identifying good readers, whereas the older/better readers could often produce a combination of reasons. The ability to produce a sophisticated rationale for a

Table 6-9. Attention Verbalization Items: Frequencies of Scores

		A-V-1	A-V-2	A-V-3	A-V-4	A-V-5	A-V-6	A-V-7	A-V-8
Grade 3									
Poor	0	6	2	9	7	12	4	4	5
	1	18	22	15	15	11	8	4	10
	2	0	0	0	2	1	12	16	9
Average	0	3	2	4	2	4	2	0	6
	1	19	22	18	11	16	12	8	9
	2	2	0	2	11	4	10	16	9
Good	0	2	0	2	1	4	1	0	7
	1	20	24	21	14	14	3	7	7
	2	2	0	1	9	6	20	17	10
Grade 6									
Poor	0	1	0	1	1	2	0	0	6
	1	15	23	20	3	19	6	5	3
	2	8	1	3	20	3	18	19	15
Average	0	0	0	0	2	0	0	0	6
	1	12	24	23	5	16	4	3	5
	2	12	0	1	17	8	20	21	13
Good	0	0	0	0	0	1	0	1	4
	1	6	23	15	1	15	1	1	7
	2	18	1	9	23	8	23	22	13
Chi square		57.48***	12.17	45.11***	58.08***	34.51**	32.15**	24.35*	8.49

*$p < .01$ **$p < .001$ ***$p < .0001$

choice increased with grade and with reading ability. In particular, the ability to produce a sophisticated response was characteristic of sixth-grade good readers more than it was of sixth-grade average readers.

The following items were designed to assess the child's knowledge of relevant cues or information in specific situations.

Question A-V-4: Suppose that you wanted to put words in alphabetical order. How would you do it?

Increasing knowledge about the use of cues was evident in this item. The younger/poorer readers either had no idea or would look only at the first or second letter. The older/better readers, however, realized that it was often necessary to look beyond the second letter. The ability to produce a sophisticated response to this item increased with grade and also from the third-grade poor to average level, the third-grade good to sixth-grade poor level, and the sixth-grade average to good level.

Question A-V-5: Suppose I gave you a list of words and asked you to put all the rhyming words together in their proper groups. How would you do it?

When explaining how to identify rhyming words, the younger/poorer readers either had no idea or, at best, knew that they should match the final letters in the words. Matching final letters was a frequently used strategy for older/better readers, but these children also often suggested that they would have to listen to the sound at the end of the words in order to be absolutely sure. The ability to produce a sophisticated response to this item increased with reading ability. This increase was particularly evident between poor and average readers.

The following items were used to assess the child's awareness of where attention had been directed during a task, and how differences in the tasks affected the difficulty of the task.

Question A-V-6 and Question A-V-7: How did you find so many [target letters] so quickly?

The younger/poorer readers had a tendency to describe the pattern of their search as "just looking" or not knowing at all. In contrast, the olders/better reader tended to explain the search pattern as systematic (e.g., down each row) and reported using redundant cues when they were present. The ability to produce a sophisticated response increased with grade for both the redundant-information task and the nonredundant-information task. In addition, on the redundant-information task, there was an increase with reading ability. This effect of reading ability largely was evident in the increase between average and good readers, particularly at the third-grade level.

Question A-V-8: Which one was easier [referring to the above task]? What made it easier?

Most of the children realized that the search task in which they could use redundant information was easier, and most could explain why. However, the ability to produce a sophisticated response to this item did increase with grade.

The relationship between each verbalization item and reading ability at each grade level also was assessed by means of correlations and partial correlations, controlling for nonverbal IQ (Appendix N, Table N-13). Once again, some of the items are related to reading ability at each grade, and this relationship is affected minimally by the removal of variance attributed to nonverbal IQ.

In addition, all of the verbalization items were combined to produce a computed score to indicate each child's level of achievement on attention verbalization. This measure, presented in Table 7-1 in a later chapter, increased with grade and with reading ability.

Overall, the younger/poorer readers showed limited knowledge about possible cues that could be used in search problems, and they rarely indicated systematic search. For example, many third-grade readers suggested that they would choose a book from the library by "looking at the cover." In contrast, one sixth-grade good reader, in response to the same question, said, "Well, sometimes I go by the cover because I can tell a lot by that, and if I can't, then I get the one, like I look at the inside cover and decide from that. It tells about the story, like part of the story, and sometimes I flip through the book and if the writing is really very small, well, I don't really like to read those kind of books, like I like fairly easy books to read, not easy readers but nice thick books, so I pick one, then I can always get it later. [Q: Anything else?] I might just start reading a bit of the first chapter to see what it is like. I can tell by that and if I don't like it, I just close the book and get another one."

As expected, the older/better reader was more able to make appropriate verbalizations about his or her attention skills. To a certain extent, this difference was a grade difference, but within each grade different items were related to reading ability. At the lower grade, the important factor seemed to involve identifying cues within words (alphabetical order, rhyming words), whereas at the sixth-grade level, the important factor involved generating alternative cues to use for identification purposes (library book, ways to tell a good reader). Once again, it must not be overlooked that the younger/poorer readers did have some limited knowledge about their attention skills. For example, most children could think of at least one way to identify a good reader and could identify the easier of two search tasks. However, it was only the older/better reader who was capable of producing sophisticated answers indicating strategic use of cues.

In addition to the quantitative differences substantiated by statistical analysis, there are qualitative differences among children at the various levels that can only be seen by actually reading the verbatim transcripts. For this reason, excerpts from selected good and poor readers at each grade will be presented next.

Language Transcripts

Overall, the younger/poorer reader seems less able than the older/better reader to express knowledge about words and sentences. However, even within this younger/poorer group, there is considerable variability. Consider the following protocols from *third-grade poor readers*. (The questions are abbreviated for space reasons, but the child's answers are reported verbatim.)

Q. Words and sentences. What can you tell me about words?
A. *No reply.*
Q. Where do we use words?
A. *No reply.*
Q. Do you use words?
A. Yeah.
Q. Where do you use words?
A. To read and to write.
Q. Can you tell me anything else about words?
A. That's it.
Q. What can you tell me about sentences?
A. *No reply.*
Q. When do we use sentences? Where do we use sentences?

A. On the blackboard, to write down.
Q. Johnny. *Tdet.*
A. Wrong. Because it's not a word.
Q. What makes it not a word?
A. *No reply.*
Q. *Meff.*
A. Right.
Q. What makes that a word?
A. *No reply.*
Q. *Stone.*
A. Right. Uh, when we're outside we pick stones up, and sometimes we have collections of stones.
Q. Which one was his best word?
A. *Stone.* Because it's a word.
Q. Is *meff* a word?
A. No.
Q. No? Should you mark *meff* wrong? Yes? Why would you mark it wrong?
A. I'm not sure.
Q. If you weren't sure, is there anything you could do to be sure?
A. *No reply.*
Q. Is there any way you could find out whether to mark it right or wrong?
A. Um-hm.

As can be seen, this particular third-grade poor reader could think of very little to say about words and sentences other than that we use them to read and write. When judging whether a group of letters was a word or not, there was some confusion as to whether or not a pronounceable nonsense syllable was or was not a word, but the child could never explain this confusion or justify the answers given. In addition, this child could think of nothing to do in order to solve the problem.

Unfortunately, the data from the previous child contained a large number of "no responses" to the probing. If this was the case with all poor readers, we might wonder if the difficulty was a lack of ability regarding general verbal responses rather than a lack of ability to understand. However, this did not seem to be the case. Many poor readers responded to the interview items in a way that suggested that they had developed coping strategies rather than metacognitive skills. For example, another *third-grade poor reader* responded in the following way.

Q. What can you tell me about words?
A. They have . . . like they can be compound words. And they learn you how to read and write. And, um . . .
Q. Where do we use words?
A. In sentences.
Q. What can you tell me about sentences?

A. They, uh . . . like they are, uh . . . there are lots of words in them. And they make a great big sentence.
Q. Where do we use sentences?
A. In reading.
Q. Johnny. *Tdet.*
A. Right. It's a French word.
Q. It's a French word? OK. How do you know, or do you just think it is?
A. I think, um, like lots of French people say it.
Q. *Meff.*
A. Right. Like, you can meff something.
Q. How do you meff something?
A. With a magnet.
Q. Mm-hm. How do you do that?
A. You get a magnet and you, uh, put the . . . get a heavy steel—steel—then the magnet touches the steel.
Q. Mm-hm. And that's meffing, is it?
A. Yeah.
Q. *Stone.*
A. That's right. Like there's a stone that you pick up, from the ground.
Q. How could you be sure how to mark his words?
A. Check 'em.
Q. How?
A. Like, uh, sound them out.
Q. So if you could sound them out, then it would be a word?
A. Yes.
Q. Which one of his words was his best word?
A. Meff.
Q. Why is meff the best word?
A. You can, like . . . it's a nice word.

This child, like the first one, did not know much about words or sentences, although he could name at least one type of word. He also felt that sentences had to be long. Presumably, his criterion for a word (whether or not you could sound it out) led him to the conclusion that both "tdet" and "meff" were words. His justifications for these judgments, however, could be the result of one of two things. Perhaps he really believed that "tdet" was a French word and "meff" had something to do with magnets, or perhaps he was simply creating a possible answer so that he could respond when a response was required. We will never know for sure which of these rationales is correct. However, it is obvious that this poor reader was quite adept at providing verbal responses when necessary. What is lacking is the knowledge that is necessary for a good answer. Another common type of response from a younger/poorer reader was to reword or restate characteristics already given. This would support the hypothesis that they have learned to give a response—of any sort—when required to do so.

In contrast, the older/better reader is more apt to provide a concise yet knowledgeable answer without much probing. For example, one *sixth-grade good reader* responded as follows.

Q. What can you tell me about words?
A. Well, all words have a vowel and most of them have consonants, um, and there are long vowels and short vowels in sentences and words, and when you have an "e" at the end, it is usually, like in the word "cake," it makes the "a" long.
Q. Where do we use words?
A. We use words all the time. Because if we didn't use words, I guess we wouldn't talk.
Q. Sentences.
A. Sentences have a subject and a predicate, and they are a complete thought. At the beginning of them, they always have a capital, and at the end, a punctuation mark.
Q. Where used?
A. We use them when we are writing and when we are talking.
Q. Johnny. *Tdet.*
A. Wrong. Because it's wrong. Because it begins with "td," and I've never seen a sentence—a word—begin like that.
Q. *Meff.*
A. Wrong. Well, I just think it's not a word. I don't know whether it is or not, it doesn't look like one, though. Because it's got two ff's at the end.
Q. *Stone.*
A. Right. Because that is how you spell stone. Because I've been taught that that is a word.
Q. Suppose you weren't sure.
A. Check with the teacher. Maybe look it up in the dictionary.

As can be seen, the older/better reader is not universally knowledgeable. Some of the answers given in the above protocol could be improved upon, but generally the quality of the responses is higher than in the case of the younger/poorer reader.

The ability to produce appropriate justification for judgments of grammatical acceptability also increased with grade and with reading ability. As will be seen in the following examples, the level of response did vary with the degree of difficulty of the sentence. In addition, the level of response was not solely dependent on age, as often the third-grade good readers could produce appropriate judgments and justifications, whereas sixth-grade poor readers often had difficulty.

Sentence #1:

Q. John park to went. Right or wrong? [If judged wrong] What would you do to make it a good sentence?

Grade 3 Poor Reader

A. I would do it wrong.

Q. Why?

A. That's not very much of a sentence.

Q. How do you know?

A. Because it's not long enough to be a sentence. John went to the park.

Grade 3 Good Reader

A. It's all mixed up and it doesn't make sense. John went to the park, I'd have to add "the."

Grade 6 Poor Reader

A. Wrong. Because it don't sound right. John went to park. John went to the park.

Grade 6 Good Reader

A. Wrong. Well, it doesn't make sense. It's a sentence, but it's not in complete order. It's all mixed up. It should be "John went." Well, there should be a "the" in there. John went to the park.

Sentence #2:

Q. My favorite toothpaste is Crest. Right or wrong? Why? [If judged wrong] What would you do to make it a good sentence?

Grade 3 Poor Reader

A. Wrong. It's not very . . . it's not kind of a sentence.

Q. How do you know?

A. Well, you would have to add more to it.

Q. What would you add to make that a good sentence?

A. My favorite toothpaste is Crest, and I brush with it.

Grade 3 Good Reader

A. Right. It makes sense.

Grade 6 Poor Reader

A. Right. 'Cause it sounds better.

Grade 6 Good Reader

A. Yeah, I'd mark that right. Because it makes sense and he's got all the capitals in, and he hasn't got anything mixed up.

Sentence #3:

Q. After school Bill wented home. Right or wrong? Why? [If judged wrong] What would you do to make it a good sentence?

Grade 3 Poor Reader

A. I would mark it right.

Q. Why?

A. 'Cause it's a long sentence.

Grade 3 Good Reader

A. I'd mark it wrong. Because there shouldn't be an "ed" at the end of "went." Because it doesn't make sense to have the "ed" on the end of that. After school Bill went home.

Grade 6 Poor Reader

A. Wrong. Because it doesn't tell where he went.

Q. How would you make that a good sentence?

A. Don't know.

Grade 6 Poor Reader

A. No. Because the "ed" is, um . . . well you don't use "ed" in a sentence. I don't think . . . "went" . . . you wouldn't add "ed" for anything. After school Bill went home.

Sentence #4:

Q. Before Mary could enter the contest. Right or wrong? Why? [If judged wrong] What would you do to make it a good sentence?

Grade 3 Poor Reader

A. I'd mark it right.

Q. Why?

A. 'Cause it's a very long sentence.

Q. How do you know it's a sentence?

A. 'Cause it's long.

Grade 3 Good Reader

A. Yes, it makes sense.

Grade 6 Poor Reader

A. Right. Sounds best.

Grade 6 Good Reader

A. No, because it doesn't have an ending. It's just, like, I guess he might have got cut off or something 'cause it doesn't have a proper ending. Before Mary could enter the contest, it was over.

Sentence #5:

Q. My favorite TV program are *Gunsmoke*. Right or wrong? Why? [If judged wrong] What would you do to make it a good sentence?

Grade 3 Poor Reader

A. Right.

Q. OK. What makes that one right? Or, it just is?

A. Just is.

Grade 3 Good Reader

A. Wrong. Because it doesn't make sense, it should be "is" *Gunsmoke*.

Q. Why "is"?

A. Well, "are" makes it says—will make it—my favourite TV programs are . . .

Q. OK. So then it should be "programs are."

A. Yeah.

Grade 6 Poor Reader

A. Wrong. Because it don't sound right either. My favorite TV program is *Gunsmoke*.

Grade 6 Good Reader

A. Wrong. Because it says TV program, and that means only one, and he wrote "are," and that would mean more than one. So, if he said "is," it would be right. My favorite TV program is *Gunsmoke*.

The young/poor reader also has less knowledge about the flexibility of language. One *third-grade poor reader* responded to questions about transformation as follows.

Q. Can you always say the same thing in different ways?

A. No.

Q. How come you have been able to give me two meanings for each of those words [homonyms]?

A. Well, there is different meanings for them and they're spelled differently, but they sound the same.

Q. Could you always give me different meanings for each word that I gave you?

A. No.

Q. Why?

A. Because some of them just have, like, have one meaning. Like they're spelled in one way and have one meaning.

It also must be noted that the younger/poorer reader appears to make more grammatical errors in conversations than does the older/better reader.

As expected, the older/better readers were more able to make appropriate verbalizations about their language skills. The older/better reader appeared to know more about words, sentences, and grammatical rules and to exhibit more knowledge about the flexible use of words. Once again, we should not assume that younger/poorer readers have no knowledge about their language skills. For example, most children were able to recognize correct sentences as correct because they make sense. In addition, they could often recognize an incorrect sentence as wrong, but could offer no explanation for their judgment. It were only older/better readers who was able to express a relatively high level of knowledge about their language skills (e.g., they could recognize an incorrect sentence as wrong and could identify and explain the actual mistake).

Memory Transcripts

Similar qualitative differences could be seen in the interview about memory skills. Overall, the third-grade poor readers seem to have little idea of their own memory processes. For example, one *third-grade poor reader* responded as follows.

Q. Telephone number. 555-8643.

A. I'd call.
Q. Why?
A. I just would.
Q. What do you do when you want to remember a phone number?
A. Well, I write it on a piece of paper.
Q. Anything else that you can do, or is that it?
A. Yeah.
Q. Do you remember any of the number that I said?
A. 5-O . . . that's it.
Q. Pictures. Harder or easier to remember with story?
A. Easier.
Q. Why?
A. *No response.*
Q. Or you just think it would?
A. Yes.
Q. Can you tell me what the pictures were?
A. A bed, best shirt, and best shoes, a table, a dog, a hat, and a door.
Q. Pictures. Three minutes to learn.
A. By looking at them carefully.
Q. Why that way? Any idea?
A. No.
Q. Is there anything else you can do?
A. Well, I can read the words.
Q. Do you learn everything that way?
A. Yes.
Q. Do you remember the pictures?
A. Carrot, grape, brown shoes . . . that's it.
Q. Finding a lost jacket.
A. By looking for it with my eyes.
Q. Where would you look?
A. The sidewalk or on the driveway.
Q. Any other place? Why the sidewalk or driveway? You just think it might be there?
A. Yes.
Q. Anything else?
A. No.
Q. Who would remember the names of people at a birthday party, the child who went home or the one who went to a play?
A. The one who went home.
Q. Why?
A. *No reply.*
Q. Any idea?
A. No.
Q. Names?

A. Cindy, Fred, that's it.

As can be seen, this third-grade poor reader has little idea of the cues that might help in memory tasks, and often can provide no reason for using memory strategies. Yet some items are remembered from each recall task. However, the *third-grade good reader* often is more able to provide adequate answers to the same questions. One example follows.

Q. Telephone number. 555-8646.
A. Right away, because I could forget easily.
Q. What do you do when you want to remember a telephone number?
A. You might write it down or keep it in your head so you really remember it.
Q. How would you keep it in your head?
A. Well, like if you went to phone someone about an hour after he told you, you just think of it and you try to remember all the numbers that are in, and then get them in the right places.
Q. Mm-hm. So you just try and think of it? Anything else?
A. That's it.
Q. Pictures. Harder or easier to remember with story?
A. Harder. Because you can't always remember the story and think. Because when you look at the pictures, like you might think, forget what was meant in the story.
Q. So you might remember other things besides the pictures?
A. Yes.
Q. Do you remember the pictures?
A. There was some grapes, squirrel, corn, mittens, carrot. There was a chick in there, and I think that's all.
Q. Finding a lost jacket.
A. I would go looking for the color of it, because you would probably know what color it was and it would probably—sometimes if you . . . you would probably put your name on it.
Q. Anything else?
A. I think that's what I would do. Just that.
Q. You'd look for something that was the right color?
A. Yeah.
Q. Would you look in any special place, or would you just look?
A. I would probably look where I left it, the last place where I had it.
Q. How could you tell?
A. Because you could remember where you were playing and probably when you were taking it off and left it then when the bell rang. That's about it.
Q. Remembering the names of people at a birthday party.
A. The one that went straight home. Because when you do—um, at the practice of the play, then you'll probably forget a little bit because you

are concentrating on the thing that you are doing and you forget the people's names.

Q. Names?

A. Fred, Sally, Anthony, Jane, Bill.

As can be seen, this third-grade good reader, with a little prompting, was able to give adequate answers to memory problems. For example, he realized that he might forget the telephone number if he gets a drink of water, and he knew how to search for lost items. Unfortunately, often the *sixth-grade poor reader* responded in a similar, or sometimes even a less sophisticated manner. One example follows.

Q. Telephone number. 555-8643.

A. I would get a drink of water first. Because like, she said like, when she comes, maybe she comes back and it could be any time like, right?

Q. Mm-hm. Somebody told you the telephone number and you wanted to call. If it doesn't need to be right away, then you wouldn't call right away?

A. No.

Q. What do you do when you want to remember a telephone number?

A. I write it down. That's it, I'll write it down.That's it, I'll write it down.

Q. Do you remember any of the numbers that I said?

A. No. 5-5-5—that's it—9?

Q. Pictures. Harder or easier to remember with story?

A. I would say easier.

Q. Why?

A. Because, like when you tell the story, it's about all these things.

Q. Do you remember the pictures?

A. A bed, a new shirt—does it have to be in order?

Q. No.

A. A hat, a dog, a pair of shoes. That's all I can remember.

Q. Pictures. Three minutes to learn.

A. What would I do? I would remember my mittens, that's for sure, because it's winter, and carrots—I would remember eating them on Sunday, and corn—well, I love corn, and squirrels—I go to the park every day and see them, and chickens—I haven't seen chickens before, I never saw a chicken. Shoes, I wear shoes. Grapes, I love grapes. A coat, I would remember my coat. A frog, I would remember French people.

Q. So you would remember something about each one. Why would you do it that way?

A. I don't know. There's no reason why, I don't think.

Q. Anything else?

A. That's it.

Q. Do you learn everything that way?

A. No.

Q. Do you remember the pictures?
A. Mittens, carrots, a frog, coat, shoes, corn, grapes, chicken—that's all I can remember.
Q. Finding a lost jacket.
A. I would go to the lost and found, and if it's not there, I would go to the principal and he would probably put it over the P.A. I'd go look for it.
Q. Where?
A. On the playground and in the school, where I used to go in school, and that's about it.
Q. Remembering names of people at a birthday party.
A. The one who went straight home. Because she got, um—the one that went to practice the play, like she had other things on her mind. She had to practice the play and everything.
Q. So she would forget?
A. Yeah.
Q. Names?
A. Bill, James, Fred, Sally, Marie, Anthony—that's about it.

As can be seen, this sixth-grade poor reader showed little knowledge about memory strategies. For example, this reader suggested ways of remembering each clusterable item, but thought of nothing that would reduce the task demands. In addition, when searching for a lost item, this reader could suggest relevant cues but did not indicate a systematic search plan. However, by the *sixth-grade good* reading level, answers to memory items were fairly sophisticated. One example follows.

Q. Telephone number. 555-8643.
A. No. It wouldn't make any difference. Well, I can remember pretty well.
Q. How do you keep it in your head?
A. Well, if we don't have any paper around, most kids can write it down on their hand. The back of the palms. I just remember it. I concentrate on it for a little while.
Q. How do you remember it?
A. I just keep on thinking of it. I forget about it and it usually comes back.
Q. Do you remember any of the numbers that I said?
A. Mm-hm. 555-8603.
Q. Pictures. Harder or easier to remember with story?
A. I think the story would help.
Q. Why?
A. Well, except in one case. I don't know if that's clothes or shirt. That could be clothes or shirt. Well, she can hear the words and go through and she can—maybe if she likes the story, it could stick in her mind.
Q. And then if the story would stick in her mind, it would make the pictures easier to remember?
A. Mm-hm.

Q. Do you remember the pictures?
A. A bed, a shirt, shoes, a table, a dog, a hat, and a door. I'm good at remembering.
Q. Pictures. Three minutes to learn.
A. Do I have to remember both yellow corn or just corn?
Q. Just corn. What would you do to learn these pictures?
A. Well, put these together, clothing that could be worn, and the animals.
Q. OK. Why would you do it that way?
A. Well, then they'd be in categories.
Q. That would be easier?
A. Mm-hm.
Q. Anything else?
A. Well, I could put 'em under letters. Like corn, carrots, coat, and chick. That wouldn't be as good as the other ones: squirrel and shoes. Mittens, frog, and grapes would be all separate.
Q. Anything else?
A. Well, they could be put in order according to colors. Yellow, purple, brown, orange.
Q. But clothing and food and animals would be best?
A. Mm-hm.
Q. Do you learn everything that way?
A. No.
Q. Do you remember the pictures?
A. Well, there's corn, grapes, carrots—so far so good—a squirrel, a chick, a frog, a coat, mittens, shoes. I got 'em.
Q. Finding a lost jacket.
A. I'd go to the lost and found, inquire with the teachers and students if they'd seen it, and I'd go back myself and look around, to where I thought I'd left it. Where I could try and retrace my steps. That's it.
Q. Remembering the names of people at a birthday party.
A. The one that went straight home. Because he didn't get in contact with that many, and the other guy got a whole bunch more. So he would know who.
Q. Names?
A. Anthony, Sally, Lois, Jim, Jane. That's about all.

As expected, the older/better reader was able to make appropriate verbalizations about his memory processes. Moreover, the older/better reader seemed to be able to identify the possible strategies and could choose among them to optimize efficiency. For example, this particular sixth-grade good reader knew that there were various ways in which to cluster pictures, but one was more efficient than the others because it included all items. In addition, this reader also could devise a strategic plan to find a lost jacket. These were skills that were lacking in the younger/poorer reader.

Use of Important Information: Attention Transcripts

Again, similar qualitative changes were seen in the interviews about use of important information. The younger/poorer readers showed limited knowledge about the possible cues that one could use in search problems and rarely indicated systematic search. For example, one *third-grade poor reader* responded as follows.

Q. Library books. Choosing one to read.
A. By looking at the cover.
Q. Anything else?
A. I would look at all the covers, too. That's about it.
Q. Deciding who is a good reader.
A. By listening to them all.
Q. Why that way?
A. Because maybe I won't find any other way.
Q. Listen for what?
A. If I don't listen to them, then I might not know who is the best reader.
Q. What would you listen for? How could you tell?
A. By listening to them. Good expression. That's about it.
Q. Any other way to tell?
A. No.
Q. Telling alphabetical order.
A. Like, look for the first letter. The first letter would have to be "a."
Q. "A"? To come first? Anything else?
A. "B."
Q. Anything else besides first letter?
A. Always look at the first letter.
Q. Telling rhyming words.
A. I would just know.
Q. Redundant-search task. How did you find them?
A. Because I just went through and see if there is a B, L, and S [target letters].
Q. Look any special way, or just look?
A. I just looked at here and went through each row.
Q. Nonredundant-search task. How did you find them?
A. I went the same way.
Q. Which one was easier?
A. The second one.
Q. Why?
A. Because A, G, and M [target letters] were easy.
Q. Easier letters to find?
A. Yes.

As can be seen, this third-grade poor reader had little idea of the relevant cues that would aid in any search. For example, the only cue that this reader was aware of that might help in picking out a good book was the cover. In addition, this child showed little knowledge of how to use cues in search situations. However, by the third-grade good reading level, many children are showing more awareness of important cues. For example, one *third-grade good reader* responded as follows.

Q. Library books. Choosing one to read.
A. Well, about dogs, or. . . .
Q. Suppose you had a whole shelf full of dog books that you might like. How would you decide which one to read?
A. Well, like if we were doing a speech or something, one that had most things that tells you about dogs.
Q. How could you tell?
A. Well, you could maybe just look at it. Maybe the title would say something about it.
Q. Would you look at anything else?
A. You could maybe look at what's inside, or maybe read a page or something like that.
Q. Anything else?
A. Well, you could maybe look at the pictures.
Q. Deciding who is a good reader.
A. Well, you could get a story—like out of the reader—and ask them to read it.
Q. Why would that be the best way?
A. Because then you can hear them out loud and see if they have any problems.
Q. What would you listen for?
A. If they didn't know the words or if they got stuck.
Q. Anything else?
A. That's about it.
Q. Telling alphabetical order.
A. Well, you first get the words that start with "a" and "b" and all the way up, and if, uh, like if two words started with "a," then you would look at the second letter, and if . . .
Q. Is there anything else you look at?
A. Well, if those two letters were the same, you'd look at the third letter.
Q. Telling rhyming words.
A. Hm, you could, uh . . . maybe think of some of your own rhyming words and say them fast and say, like, these other two words, see if they rhyme.
Q. How could you tell if they rhymed?
A. Well, like I was going to say something, but now I forgot it.
Q. Any idea how you could tell if they rhymed, or would you just know?
A. You'd just know.

Q. Redundancy-search task. How did you find them?
A. I just—I went down the lines.
Q. Nonredundant-search task. How did you find them?
A. Well, I just looked across and then down. Like I went down each row like that.
Q. And how did you go across?
A. Over like that.
Q. So you did one and then the other, or did you do both at the same time?
A. The one and then the other.
Q. Which was easier?
A. The first one. Well, it just was easier.

As can be seen, this third-grade good reader was aware of many more possible cues than the previous poor reader. In fact, there are some indications that this good reader was beginning to use cues strategically. For example, this reader provided a fairly accurate plan for arranging words in alphabetical order and could explain the strategy used in the search task. However, this child did not seem to be aware of all cues. For example, she did not realize why the redundant-search task was easier.

Even though most children have a reasonable amount of knowledge of important cues, by the time they reach sixth grade, some still have difficulty using these cues strategically. In fact, it is difficult to differentiate some sixth-grade poor readers from third-grade children on the basis of their responses. For example, one *sixth-grade poor reader* responded as follows.

Q. Library books. Choosing one to read.
A. You look for the exciting one.
Q. How could you tell?
A. By looking at the cover, the pages, and you read part of it, and then you find out how interesting it is.
Q. Deciding who is good reader.
A. I don't know. The one that talks, reads more louder, and reads it clear. The way they say it, the way they express it.
Q. Telling alphabetical order.
A. Look at the first letter and if they're the same you go to the second, and then you go . . . then you . . . then you just put it in alphabetical order by the letters.
Q. Ever look at anything else besides the first two letters?
A. That's all I do.
Q. Telling rhyming words.
A. See if they sound the same.
Q. So you would say them out loud, or to yourself?
A. To yourself.
Q. Redundant-search task. How did you find them?

A. Go like this [indicated up and down rows] and keep saying b,l,s,b,l,s [target letters] and try to remember.

Q. You go down each one of the rows? OK. That's great. Nonredundant-search task. How did you find them?

A. Just keep in mind what you are looking for.

Q. How did you look?

A. Yeah, went up and down.

Q. Which was easier?

A. They were both easy, the same thing you look for.

In spite of the fact that this sixth-grade poor reader was responding at a fairly low level, many sixth-grade children had a fairly good grasp of important cues and strategic use of these cues. In fact, the following *sixth-grade good reader* is an example of the high level of response that many sixth-grade children approximated.

Q. Library books. Choosing one to read.

A. Well, usually I take a whole bunch of books from the library at once. I read all of them.

Q. How would you pick out one?

A. Well, sometimes I go by the cover because I can tell a lot by that, and if I can't, then I get the one, like I look at the inside cover and decide from that. It tells about the story, like part of the story, and sometimes I flip through the book and if the writing is really very small, well I don't really like to read those kind of books, like I like fairly easy books to read, not easy readers, but nice thick books, so I pick one, then I can always get it later.

Q. Anything else?

A. I might just start reading a bit of the first chapter to see what it is like. I can tell by that, and if I don't like it, I just close the book and get another one.

Q. Deciding who is a good reader.

A. Well, I should know already if they've been at school for a while. I could let each person have a turn at reading part of the story, or whatever we were doing at the time. I would let them have a certain paragraph and then let them sit down so that everyone will have a turn and I can judge by that.

Q. Why the best way?

A. Well, I would be hearing them themselves and they wouldn't even know that I was kind of testing them, because if I was, they would be really careful and would want to make a good impression.

Q. What would you listen for?

A. If they know how to say the words and pronunciation, like if they put fury into it, like a lot of expression and not read too fast or too slow.

Q. Anything else?

A. I could look at their work because questions asking about the story—If they are good readers, then they should have good answers if they know what the story really is about, or if they're not, then they are not very good answers.
Q. Telling alphabetical order.
A. Well, I would look for all the words starting with "a." If there is a whole bunch of those, I just get the ones that had the next letter, "ab" or "ac," or whichever came next and so on.
Q. Is there anything you would do besides look at the first two letters?
A. Well, if the first two letters were exactly the same, I would go on to the third or fourth.
Q. Telling rhyming words.
A. I could tell by the end of the word, look at the ends, and most . . . like the ones that rhymed, they have the same endings or they sound the same. I could sound them out.
Q. Redundant-search task. How did you find them?
A. Well, the first ones I found were in squares. Like I wasn't sure at first, and when I couldn't find any that didn't have circles or squares around them, I just looked in all the squares and that's where they all were.
Q. Look any special way?
A. I just looked all over the place. Well, first I was going across one row and down each row.
Q. Nonredundant-search task. How did you find them?
A. Well, I just went in order—all down the rows, that's the way I found them. And then I just looked over, across, and any way just to see if I could find any. That's why the end ones are a lot later.
Q. Which easier?
A. First one.
Q. Why?
A. After I found out they were all in squares, like all the squares were there and looked in all the squares, it lowered the number.

Once again, it must be noted that the sixth-grade reader seems to be able to think of many relevant cues that could be used in any search problem. For example, this reader knew the cues and the search strategy used in selecting a good book and showed use of relevant cues in the search tasks. In addition, their thoughts seem to be more complete; points are strung together in a logical sequence and there is less need for prompting from the interviewer.

As expected, the older/better readers were more able to make appropriate verbalizations about their attention skills. To a certain extent, this difference was a grade difference, but within each grade different items were related to reading ability. At the lower grade, the important factor seemed to involve identifying cues within words (alphabetical order, rhyming words), whereas at the sixth-grade level, the important factor involved generating alternative cues to use for identification purposes (library book, ways to tell a good reader). Once again, one must not forget that the younger/poorer readers do have some limited knowledge about their attention skills. For example, most children could think of at least one way to identify a good reader and could identify the easier of two search tasks. However, it was only the older/better reader who was capable of producing sophisticated answers indicating strategic use of cues.

Relationship Between Performance and Verbalization

The two computed scores, one for language performance and one for language verbalization, were entered into a stepwise multiple-regression equation to predict reading ability at each grade. Nonverbal IQ also was included in the equation. At the third-grade level, the multiple-regression equation accounted for 49.5% of the variance (R = .7032). Language performance accounted for 41.09% of the variance, language verbalization accounted for an additional 7.30% of the variance, and nonverbal IQ accounted for only an additional 1.07% of the variance. At the sixth-grade level, the multiple-regression equation accounted for 61.47% of the variance (R = .7840). Language performance accounted for 53.85% of the variance, whereas language verbalization accounted for only 0.14% of the variance. In a predictive sense, language performance was more useful than language verbalization. In addition, nonverbal IQ accounted for relatively litte additional variance.

The two computed scores for memory performance and memory verbalization, along with nonverbal IQ, likewise were entered into a stepwise multiple-regression equation to predict reading ability at each grade level. At the third-grade level, the multiple-regresion equation accounted for 27.16% of the variance (R = .5211). Memory performance accounted for a 6.50% of the variance, and memory verbalization accounted for only an additional 0.12% of the variance. At the sixth-grade level, the multiple-regression equation accounted for 41.15% of the variance (R = .6415). Nonverbal IQ accounted for 32.88% of the variance, memory performance accounted for an additional 5.04% of the variance, and memory verbalization accounted for an additional 3.23% of the variance. In a predictive sense, then, memory performance was a more powerful predictor at the third-grade than the sixth-grade level, and was more powerful at both grades than was memory verbalization.

Finally, the overall computed scores, one for attention performance and one for attention verbalization, were entered with nonverbal IQ into a stepwise multiple-regression equation to predict reading ability at each grade. At the third-grade level, the equation accounted for 39.41% of the variance (R = .6277). Attention performance accounted for 32.01% of the variance, attention verbalization accounted for an additional 5.83% of the variance, and nonverbal IQ accounted for only an additional 1.57% of the variance. At the sixth-grade level, however, the pattern of results was somewhat different. The regression equation at this level accounted for 43.45% of the variance (R = .6592). Nonverbal IQ was the best predictor, accounting for 32.88% of the variance, followed by attention performance, which accounted for an additional 7.50% of the variance. Attention verbalization accounted for only an additional 3.07% of the variance. It appears that the most important changes in attentional skills are between, rather than within, grade. Within each grade, performance measures appear to be much more useful than verbalization measures in predicting reading ability. However, even the performance measures may not be as useful as nonverbal IQ as a predictor, particulary at the higher grade level.

Metacognitive Categorizations

Finally, the children were classified according to the definition of "meta" described in the General Method chapter. The frequency of children falling into each language category at each reading level is presented in Table 6-10. Also shown are comparable frequencies for categorizations on memory and attention.

Language

As can be seen in Table 6-10, there was a progression from being low on both performance and verbalization to being mature metacognizers. The third-grade poor readers were classified almost exclusively as LOW, while the sixth-grade good readers were almost exclusively classified as META. The frequency of children classified as LOW declined as grade and reading ability increased until, by the sixth-grade average reading level, no children fell in this category. Concurrently, very few children below the third-grade good reading level were classified as META. In addition, very few children (less that one-sixth of all the children) were classified as MIMICS or as in TRANSITION. Comparing the number of children in less mature categories with the number of children in the META category, the number of mature metacognizers increased with grade and with reading ability. Poor readers differed from average readers, and average readers differed from good readers. In particular, third-grade average and good readers were categorized differently. (Statistical tables for these analyses appear in Appendix N.)

Memory

As shown in Table 6-10, there is a progression from being low on both performance and verbalization of reading skills to being mature metacognizers. To a certain extent, this change appears to be a function of grade rather than of reading ability. One-third of all children tested were classified as LOW, and most of

Table 6-10. Numbers of Children Categorized as LOW, MIMIC, TRANSITION, and META

Language

	Grade 3			Grade 6			
Category	Poor	Average	Good	Poor	Average	Good	%
LOW	23	13	5	8	0	0	34.0
MIMIC	0	5	2	4	2	0	9.0
TRANS.	1	3	6	1	4	3	12.6
META	0	3	11	11	18	21	44.4

Memory

	Grade 3			Grade 6			
Category	Poor	Average	Good	Poor	Average	Good	%
LOW	18	15	11	3	1	0	33.3
MIMIC	4	5	4	4	1	1	13.2
TRANS.	2	2	5	7	5	4	17.4
META	0	2	4	10	17	19	36.1

Attention

	Grade 3			Grade 6			
Category	Poor	Average	Good	Poor	Average	Good	%
LOW	22	14	12	1	0	0	34.0
MIMIC	1	3	4	0	0	0	5.6
TRANS.	1	4	3	5	6	2	14.6
META	0	3	5	18	18	22	45.8

these were from the third grade. In contrast, most of those classified as META were from the sixth-grade level. Within each grade there also was some progression of reading ability. In addition, only 13.2% of the children were classified as MIMICS and 17.4% of the children were classified as in TRANSITION. Neither category showed a distinctive pattern. Comparing the number of children in less mature categories with the number of children in the META category, we found that the number of mature metacognizers increases with grade and with reading ability.

Use of Important Information: Attention

As can be seen in Table 6-10, the change from being classified as LOW to being classified as META appears to be a function of grade. Of those falling in the LOW category, all but one (a sixth-grade poor reader) came from the third grade. Only 5% of the total number of children fell into the MIMIC category, and all were from the third grade. The low number of children in the MIMIC category is not surprising since meta-attention skills are not taught and rarely are considered, even casually, in schools. The number of children in the TRANSITION category also was low (15% of the total) and presented no particular pattern. Almost half (46%) of the children tested fell in the META category, most of them from the sixth grade. Comparing the number of children in less mature categories with the number of children in the META category, we found that the number of mature metacognizers increases with grade. In addition, the third-grade good readers differed from the sixth-grade readers on this measure. It also is interesting to note that attention differs from the other processes (language, memory) in that there appears to be a definite and sharp change in attention development rather than a gradual and continuous change with grade and reading ability.

Overview of Developmental Components

It seems clear that the older/better reader has a greater ability than the younger/poorer reader to perform on language tasks and verbalize about language skills. As for the predictive power of these two aspects of language development in relation to reading ability, the ability to perform on language tasks is much more important than the ability to give appropriate verbal responses about language skills. However, this is not to say that language verbalization skills are less important. In fact, putting it all together, only the older/better reader had both high performance and high verbalization skills. Very few children fell into the category of high performance and low verbalization skills. It also should be noted that nonverbal IQ did not prove to be a strong predictor.

Much the same is true with respect to memory. That is, one cannot dispute the fact that the older/better reader has a greater ability than the younger/poorer reader to perform on memory tasks and to verbalize about memory processes. Considering the predictive power of these two aspects of memory development in relation to reading ability, the ability to perform on memory tasks is much more important than the ability to give appropriate verbal responses about memory skills. However, even memory performance is much more impressive as a predictor of reading ability at the third-grade level than at the sixth-grade level. In spite of the fact that memory performance appears to be a good predictor, it is not necessary to conclude that memory verbalization skills are less important. In fact, when viewed together, it was found that only the older/better reader had both high performance and high verbalization skills. As mentioned before, very few children fell into the category of high performance and low verbalization skills.

It is clear also that the older (and to some extent, the better) reader has a greater ability than the younger (poorer) reader to perform on attention tasks and to verbalize about important information and search strategies. Considering the

Table 6-11. Developmental Components: Predicting Reading Ability

	Grade 3		Grade 6	
Variable	Order	Variance	Order	Variance
L-P	1	41.09%	1	53.85%
A-P	4	0.88%	6	0.27%
M-P	3	5.31%	5	0.41%
L-V	2	7.30%	7	0.01%
A-V	7	0.07%	3	2.37%
M-V	5	0.30%	4	0.56%
IQ	6	0.26%	2	7.48%
Total		55.21%		64.94%

predictive power of these two components, performance on attention tasks is much more important than the ability to verbalize about attention skills. However, the predictive power of attention performance is greater at the third-grade level than at the sixth-grade level. In spite of the fact that performance is the stronger predictor, it is not necessary to claim that knowledge about attention skills is less important, because high performance skills rarely occur without high verbalization skills. When performance and verbalization are viewed together, it is the case that the change from LOW to META is related primarily to grade rather than to reading ability. Perhaps this is (at least in part) the reason why attention, in general, failed to be a particularly strong predictor at either grade.

As will be recalled, one of the original aims of this study was to examine the relationship between various types of skills. In order to examine the relationship between each developmental component and general reading ability at each grade, the computed scores, along with nonverbal IQ, were simultaneously entered into a regression equation to predict reading ability. The computed scores that were included in these analyses were performance and verbalization measures as related to language, attention, and memory (representing Sets 7 to 12, Table 2-1). The results of these analyses are presented in Table 6-11, in the form of the order of entry into the equation and the amount of incremental variance accounted for by each variable.

At the third-grade level, the multiple-regression equation accounted for 55.21% of the variance (R = .7431), with language performance being the best predictor. Language verbalization and memory performance also appear to be relatively important as predictors. At the sixth-grade level, the multiple-regresion equation accounted for 64.94% of the variance (R = .8059). Once again, language performance appeared to be the best predictor. As can be seen, language performance appears to be a consistant predictor of reading ability at both grades. Language verbalization and memory performance also appear to be relatively important as predictors at the third-grade level, but not at the later grade.

It appears, then, that language skills have a strong link to reading ability at both grades, but that developmental changes to component processes mainly affect the lower grade. It is probable that most developmental changes have occurred by the sixth-grade level. In addition, since both attention and memory seemed to be strongly determined by grade level, it is possible that these component processes may be prerequisites for mature reading skills. For example, it can be argued that mature reading skills depend to a large degree on one's ability to direct attention to important units and retain this information for later use through the use of efficient mnemonic strategies.

CHAPTER 7

Computed Scores: Results and Discussion

Some of our readers may wonder why we have relied heavily on computed scores rather than on item scores for our analyses, as many others have done in the past. In reality, we have very little information about cognitive–metacognitive connections, and this generally is limited to the area of memory (e.g., Borkowski, Reid & Kurtz, in press; Yussen & Berman, 1981). Moreover, the results are not always encouraging. For example, Cavanaugh and Borkowski (1980) assessed metamemory (through interview items developed by Kreutzer, Leonard & Flavell, 1975) and memory performance in children in kindergarten and grades one, three, and five. They reported significant correlations between interview and performance items when data were combined across grades, but within-grade correlations were not significant and did not generalize across memory tasks.

On the whole, the contention that successful metamemory is a necessary prerequisite for successful memory was not supported in the Cavanaugh and Borkowski study. However, it is axiomatic in the world of test construction that individual items typically are less reliable and have less predictive validity than a composite score based on a set of related items. It is probable that a similar phenomenon is affecting the data presented by Cavanaugh and Borkowski. If a composite score of memory performance was correlated with a composite score of memory knowledge, then the possibility of finding cognitive–metacognitive relations might improve. (See Rushton, Brainerd & Pressley, 1983, for an especially indicting commentary on the use of individual items in research on children's behaviors.)

In our study, we in fact did base many analyses on composite scores of performance and verbalization. The correlations between performance and verbalization composite scores within each grade were significant for all the skills assessed, with the exception of decoding. The mean correlation between performance and verbalization scores for all of the processes involved (e.g., language performance–language verbalization, decoding performance–decoding verbalization) was .41 for the third-grade children and .46 for the sixth-grade children. Also, composite scores were much better predictors than item scores of the relationships between the different processes and reading ability. The mean correlations between reading ability and the performance items was .40 for the third-grade children and .38 for the sixth-grade children. However, when the composite scores were used, the mean correlations for reading ability and performance jumped to .62 and .66 for third- and sixth-grade children, respectively. This difference between item and composite scores was even more important in the verbalization measures. The mean correlations between reading ability and verbalization items was .15 and .18 for third- and sixth-grade children, respectively. Both correlations are nonsignificant. However, when composite scores were used, the mean correlations between reading ability and verbalization jumped to a respectable level of significance, .37 and .42 for third- and sixth-grade children, respectively.

As a result of the strength of the results and analyses based on computed scores, we feel that it is worthwhile to devote a reasonable amount of time to discussion of the various analyses. As our readers will remember, an original intent of this study was to examine the relationship between each type of reading skill and each component skill. For each child, the 12 computed scores included in these analyses were both the performance and verbalization scores for decoding, comprehension, and strategies (efficiency), and for language, attention, and memory. The point of these analyses is simply to provide an overview of the pattern of relationships among the skills and of changes with grade and with reading ability.

Relationship of Computed Scores to Grade and Reading Ability

Means for all 12 of the computed scores are presented in Table 7-1. These scores and nonverbal IQ were entered into a grade X reading ability multivariate analysis of variance. The multivariate analysis of variance showed a main effect of grade, $F(13, 126) = 51.90$, $p < .0001$, and a main effect of reading ability $F(26, 250) = 11.54$, $p < .0001$. In addition, the grade X reading ability interaction was significant, $F(26, 250) = 1.75$, $p < .05$.

All of the univariate F's have been presented in the appropriate results sections. As will be recalled, all of the scores, except for nonverbal IQ, increased with grade. With only a few exceptions, the computed scores also differentiated among the three reading groups (i.e., good readers were better than average readers, and average readers were better than poor readers). Decoding verbalization was the only computed score that failed to increase with reading ability. Among the exceptions, memory verbalization only differentiated poor readers from good readers, and memory performance only differentiated poor readers from average and good readers. In addition, attention performance showed an interaction

Table 7-1. Computed Scores: Means

Measure	Grade 3 Poor	Grade 3 Average	Grade 3 Good	Grade 6 Poor	Grade 6 Average	Grade 6 Good
Performance						
Decoding	8.71	9.39	10.00	9.99	10.61	11.30
Compre.	9.22	9.91	10.25	9.75	10.19	10.69
Strat (Eff)	9.15	9.83	10.21	9.70	10.22	10.90
Language	9.20	9.78	10.12	9.99	10.33	10.59
Attention	9.25	9.60	9.86	10.33	10.39	10.57
Memory	9.15	9.64	9.71	10.23	10.53	10.74
Verbalization						
Decoding	9.59	9.72	9.88	10.23	10.23	10.36
Compre.	9.31	9.68	10.10	10.02	10.37	10.53
Strategies	9.39	9.63	9.85	10.17	10.32	10.65
Language	9.45	9.78	9.97	10.08	10.26	10.47
Attention	9.34	9.76	9.92	10.17	10.27	10.54
Memory	9.64	9.74	9.82	10.06	10.28	10.47

between grade and reading ability. At the third-grade level, poor readers scored lower on attention performance than average readers, and average readers scored lower than good readers. At the sixth-grade level, poor readers scored lower on attention performance than good readers.

In addition to the analyses of variance, the relationship between each computed score and reading ability at each grade level was assessed by means of correlations and partial correlations, controlling for nonverbal IQ. These correlations are reported in Table 7-2. Performance variables generally were more highly correlated with reading ability than were their verbalization counterparts. When nonverbal IQ was statistically controlled, memory verbalization and strategies verbalization failed to show a significant relationship with reading ability at the third-grade level, and decoding verbalization failed to show a relationship at the sixth-grade level. (The complete correlational matrices are presented in Appendix U.)

It appears, then, that all reading skills and component skills measured in this study (with the exception of decoding verbalization) did increase with grade and with reading ability. As has been noted, decoding strategies were actively taught in the school system, and poor readers are quite capable of using mimicking as a coping strategy. Indeed, as shown in the decoding analyses reported earlier, a high proportion of poor readers were classified as MIMICS. It also should be noted that the memory scores, both verbalization and performance, failed to differentiate among all three reading groups. It is possible that memory is a process that is associated, to a large extent, with age rather than with reading ability. If memory is a problem of reading, then it appears to be a problem

Table 7-2. Computed Scores: Correlations with Reading Ability

Measure	Grade 3		Grade 6	
	r	partial *r*	*r*	partial *r*
Performance				
Decoding	.7384***	.6960***	.7461***	.6980***
Compre.	.6082***	.5166***	.6701***	.6171***
Strat (Eff)	.7102***	.6414***	.7931***	.7353***
Language	.6410***	.5711***	.7338***	.6510***
Attention	.5658***	.4773***	.5178***	.3342**
Memory	.4532***	.3612***	.5091***	.2739**
Verbalization				
Decoding	.2338*	.2550*	.1990*	.1127
Compre.	.5207***	.4252***	.5672***	.4980***
Strategies	.3049**	.1647	.4428***	.2786**
Language	.5800***	.5063***	.4818***	.4064***
Attention	.4405***	.3462***	.4057***	.3338**
Memory	.1473	.0117	.4200***	.2567*

*$p < .05$ **$p < .01$ ***$p < .001$

mainly for the poor reader. (Memory items tended to differentiate poor readers from average and good readers.) In addition, the interaction of grade and reading ability found with attention performance might indicate the beginning of a ceiling effect. In spite of the slight variations found with decoding verbalization, memory performance, memory verbalization, and attention performance, it appears that reading skills and the component skills of language, attention, and memory all increase with grade and, to a large extent, with reading ability.

Predictive Power of Computed Scores

To further examine the overall relationship between each computed score and reading ability at each grade, the computed scores and nonverbal IQ were entered into stepwise multiple-regression equations to predict reading ability at each grade. Overall results of these regression analyses are summarized in Table 7-3, in the form of order of the entry into the equation and the amount of incremental variance.

At the third-grade level, the multiple-regression equation accounted for 72.02% of the variance ($R = .8487$). As can be seen in Table 7-3, decoding performance was the best predictor of reading ability at this grade, accounting for 54.52% of the variance. The ability to verbalize about language skills was the second best predictor, accounting for an additional 6.46% of the variance. It also should be noted that, in general, performance measures were better predictors than their verbalization counterparts. Finally, nonverbal IQ was not a powerful predictor

Table 7-3. Computed Scores: Predicting Reading Ability

Variable	Grade 3		Grade 6	
	Order	Variance	Order	Variance
D-P	1	54.52%	4	1.47%
C-P	4	2.46%	2	10.80%
S-P	3	5.77%	1	62.90%
L-P	8	0.05%	3	5.27%
A-P	—	0.00%	7	0.27%
M-P	6	1.19%	—	0.00%
D-V	12	0.01%	6	0.72%
C-V	5	0.96%	8	0.20%
S-V	10	0.03%	11	0.04%
L-V	2	6.46%	12	0.03%
A-V	11	0.01%	9	0.06%
M-V	7	0.53%	10	0.02%
IQ	9	0.03%	5	1.32%
TOTAL		72.02%		83.10%

*Did not enter equation.

when combined with these variables. This, however, does not negate the fact that nonverbal IQ and reading ability are correlated.

At the sixth-grade level, the multiple-regression equation accounted for 83.10% of the variance (R = .9116). As can be seen in Table 7-3, strategies (efficiency) performance was the best predictor of reading ability at this grade, accounting for 62.90% of the variance. Comprehension performance accounted for an additional 10.80% of the variance. In general, performance measures were better predictors than their verbalization counterparts. Once again, nonverbal IQ did not account for a substantial amount of variance.

In a predictive sense, then, decoding performance was the most useful measure at the third-grade level. However, at the sixth-grade level, the pattern of results changed and strategies (efficiency) performance was the most powerful predictor of reading ability.

Since most researchers have taken either a cognitive or a metacognitive approach—rather than a combined approach—to reading, the performance and verbalization scores were used in separate regression equations. These analyses were done for those researchers who would be interested in only these portions of the data. Considering first the performance scores alone, at the third-grade level, the equation accounted for 69.12% of the variance (R = .8314), with the best predictor being decoding. At the sixth-grade level, the equation accounted for 82.05% of the variance (R = .9058), with strategies (efficiency) being the best predictor. The order of entry into the equation and the amount of additional variance accounted for by each variable is presented in Table 7-4. The results

Table 7-4. Performance Computed Scores: Predicting Reading Ability

	Grade 3		Grade 6	
Variable	Order	Variance	Order	Variance
D-P	1	54.52%	4	1.47%
C-P	3	5.39%	2	10.80%
S-P	2	8.15%	1	62.90%
L-P	6	0.07%	3	5.27%
A-P	5	0.31%	6	0.28%
M-P	4	0.66%	—	0.00 *
IQ	7	0.02%	5	1.32%
TOTAL		69.12%		82.05%

*Did not enter equation.

of the separate performance analyses did not differ a great deal from the overall regression analyses.

Considering verbalization scores, at the third-grade level, the equation accounted for 47.08% of the variance (R = .6861), with the best predictor being language. At the sixth-grade level, the equation accounted for 57.81% of the variance (R = .7603), with the best predictor being nonverbal IQ. At the sixth-grade level, verbalization measures were not as highly correlated with reading ability as was the nonverbal IQ measure. However, comprehension did account for an additional 16.64% of the variance. The order of entry into the equation and the amount of additional variance accounted for by each variable is presented in Table 7-5. It appears that verbalization (traditional "meta") is more important at the third-grade level than at the sixth-grade level. At the third-grade level, the most important verbalization measure was language. Verbalization was im-

Table 7-5. Verbalization Computed Scores: Predicting Reading Ability

	Grade 3		Grade 6	
Variable	Order	Variance	Order	Variance
D-V	7	0.02%	6	0.09%
C-V	2	9.83%	2	16.66%
S-V	5	0.78%	5	0.78%
L-V	1	33.64%	3	4.78%
A-V	3	1.34%	4	2.63%
M-V	6	0.39%	—	0.00 *
IQ	4	1.07%	1	32.88%
TOTAL		47.08%		57.81%

*Did not enter equation.

portant, however, at the sixth-grade level, but only in connection with knowledge about monitoring comprehension.

General Discussion of Computed Scores Analyses

Overall, the younger/poorer reader tended to do less well on performance and verbalization (traditional "meta") measures in all of the areas with which this study was concerned. The results indicated that the variables related to reading ability are different at third grade and in sixth grade. At the third-grade level, a large portion of the variance in reading ability was accounted for by performance on decoding tasks. However, at the sixth-grade level, strategies (efficiency) performance was the best predictor of reading ability.

In terms of the relative importance of cognitive and metacognitive skills, measures of performance are the best predictors of reading ability. The verbalization measures alone (traditional metacognition) were relatively poor predictors of reading, especially at the sixth-grade level. If verbalization is an important measure to consider, it appears that it would be more useful at the third-grade level than at the sixth-grade level. At the sixth-grade level, correlations between verbalization measures and reading ability were not as high as the correlation between nonverbal IQ and reading ability. Language verbalization appeared to be a fairly useful predictor at the third-grade level. However, at the sixth-grade level, if verbalization is important, it appeared to be primarily so in connection with monitoring comprehension. It is possible that the sixth-grade children are still acquiring "meta" competencies that are connected with more advanced reading skills. At this point, the reader also is reminded that in each of the areas studied, *high performance rarely was found in the absence of high verbalization*.

In summary, it seems reasonable to conclude the following: (1) Both cognitive and metacognitive measures of decoding, strategies, and comprehension increase with grade and with reading ability. (2) Both cognitive and metacognitive measures of language, attention, and memory increase with grade and with reading ability. (3) Considering all the factors involved in reading, the best predictors of reading ability seem to be decoding performance at the third-grade level and strategies (efficiency) performance at the sixth-grade level. (4) Performance measures appear to be better predictors of reading ability than their verbalization counterparts, especially at the sixth-grade level. If verbalization is important, it is important in connection with language at the third-grade level and with monitoring comprehension at the sixth-grade level.

CHAPTER 8

General Discussion and Conclusions

In order to ease the memory load for our readers, we will begin this chapter with a series of brief summary statements of each of the sets of data presented earlier. As the reader will see, there is a great deal of consistency in the patterns of results. At times this leaves us with a feeling of redundancy in our writing, but we hope our readers will be patient with our repetitions and find helpful our efforts to reiterate and reinforce our point about the consistency of the results.

In general, both *performance* on each type of skill (decoding, comprehension, strategies, language, attention, and memory) and the ability to *verbalize* about each skill increased with grade and with reading ability. Performance scores tended to be better predictors of reading ability, per se, than verbalization scores. The nonverbal IQ measure accounted for very little additional variance in most cases. In addition (and not surprising, given the factors involved in this study), many of the variables were highly intercorrelated, and the results of the regression equations should be interpreted with this in mind. In spite of the fact that verbalization did not appear to be an important factor in the regression equations, we do not wish to conclude that the ability to verbalize about each skill is not important. High performance scores rarely occurred without high verbalization scores. In addition, high verbalization without high performance rarely was found, and when found, it was with younger/poorer readers. In the strictest sense, taking into account both performance and verbalization, the frequency of mature *metacognition* increased systematically with grade and with reading ability. Examples of and exceptions to this general pattern of results are noted below.

Reading Skills

Decoding

There was a tendency among sixth-grade poor readers to mimic knowledge about decoding strategies, that is, to talk about decoding strategies but not to use them effectively in a performance task. This tendency to mimic is probably the result of heavy emphasis by teachers on how to decode words. In spite of the fact that poor readers do not seem to use these skills, the ability to mimic what the teacher has said is a means by which a child copes, at least to some degree, with his practical classroom problems concerning decoding.

Comprehension and Advanced Strategies

On the reading tasks used here, it appears that children begin to use reading strategies by the third grade. At this level, good readers' comprehension scores dropped when reading for one specific piece of information (skimming). However, it was not until the sixth-grade level that good readers' comprehension increased in all other conditions (make up a title, study, read for fun), as compared to the Skim condition. In addition, analyses of covariance, controlling for nonverbal IQ, confirmed that the pattern of results just described was not attributable to nonverbal IQ. Also, the prediction-accuracy measure showed that the ability to predict comprehension accuracy increased with grade and with reading ability.

Overall Analyses of Reading Skills

In summary, then, we feel that the reading skills of decoding, comprehension, and advanced strategies of reading for a purpose all increase both cognitively and metacognitively with grade and with reading ability. The results of the regressions clearly indicate that different skills play different roles in determining performance on the Gates-MacGinitie Comprehension Subtest at different grade levels. If the Gates-MacGinitie can be assumed to represent reading ability in a more general sense, then additional speculations are possible. For example, if a researcher wanted to predict reading ability at each of the two grades, we would suggest that different skills be examined at each grade.

In addition, we feel that the subtleties of the reading skills data point to numerous possibilities for future research. For example, in the area of decoding, it would be interesting to know what strategies children were using in different situations. With regard to comprehension and advanced strategies of reading for a purpose, an important area for future research is the investigation of the monitoring process that must be involved in mature reading.

Developmental Components

Language

The older/better reader is better able both to use and to talk about language skills. The problem for younger/poorer readers seems to be an inability to recognize problems, which is a monitoring problem. Even if mistakes are recognized, the younger/poorer reader often has little idea of how to attack the situation. For example, many younger/poorer readers realized when a sentence did not "sound right," but could not identify the grammatical error or correct the mistake.

Memory

The older/better reader is better able to use appropriate memory strategies and discuss memory skills, whereas the younger/poorer reader is less apt to use effective memory strategies and gives little indication that he or she is aware of what might be done to correct a memory failure. For example, the younger/poorer reader had trouble thinking of ways to find a lost jacket, and was less likely to be able to suggest efficient strategies for remembering (e.g., clustering).

Identifying Important Information: Attention

Most of the developmental changes regarding attention have occurred by the sixth-grade level, and as a result, the importance of attentional differences is becoming minimal by this point. Attention is not a strong predictor of reading ability within grade level, and most attention changes occur between grades (e.g., as seen in the metacategorizations).

Overall Analyses of Developmental Components

Language skills have a strong link to reading ability at both grades, but developmental changes in attention and memory mainly affect performance at the lower grade. We feel that most developmental changes in attention and memory that are important in learning to read have already occurred by the sixth-grade level.

Once again, within each of the areas, the subtleties of the data point to numerous research possibilities. For example, it would be interesting to investigate the role that knowledge of grammatical acceptability plays in the monitoring of comprehension. A researcher interested in advanced reading skills would be well advised to examine in more detail the role of important units (for instance, do poor readers attend to different cues [i.e., attend to wrong cues] or simply fail to recognize that some cues are more important than others [i.e, not attend selectively])? In addition, the use of memory strategies in connection with recall and comprehension of prose should be investigated.

Combining All Factors

As we have explained above, the third-grade readers were best characterized by their decoding performance scores, whereas sixth-grade readers were best characterized by their use of advanced strategies as indicated by efficiency. In spite of the fact that most measures increased with grade and with reading ability, it should not be concluded that the younger/poorer reader has no ability to use or talk about the skills examined in this study. Rather, it should be concluded that the younger/poorer reader has less competence in each area than the older/better reader. For example, the younger/poorer readers have little difficulty recognizing a correct item as correct because it makes sense. However, the same type of reader has difficulty recognizing mistakes. If the mistake is recognized, the younger/poorer reader seems to have little idea of how to correct it. Therefore, the younger/poorer reader not only has difficulty with using any particular skill, but also with monitoring progress and correcting failures.

Relationships Between Cognition and Metacognition

One of the issues that we have avoided thus far is that of causal links between cognition and metacognition. Unfortunately, as we explained in the introduction, very little is known about the nature of the connections between cognition and metacognition. In fact, we speculated that some of the more negative positions regarding cognitive–metacognitive connections actually were based on item rather than composite scores (computed scores) and as such were psychometrically less reliable. As the reader will recall, in this study we defined metacognition as knowledge about only those skills where performance was high. However, since this definition required both performance and verbalization measures, it also was possible to examine metacognition as it traditionally has been defined (e.g., primarily in terms of verbalization measures).

On an empirical basis, we looked at the relation between performance and verbalization, using both composite and item scores. The correlations between performance and verbalization composite scores (computed scores) within each grade were significant for all the skills assessed, with the exception of decoding. The mean correlation between performance and verbalization scores for all of the processes involved (e.g., language performance–language verbalization, decoding performance–decoding verbalization) was .41 for the third-grade children and .46 for the sixth-grade children. Also, as compared to item scores, composite scores were much better predictors of the relationships between the different processes and reading ability. The mean correlations between reading ability and the individual performance items was .40 for the third-grade children and .38 for the sixth-grade children. However, when the composite scores were used, the mean correlations for reading ability and performance jumped to .62 and .66 for third- and sixth-grade children, respectively. This difference between item and composite scores was even more important for the verbalization measures. The mean correlations between reading ability and individual verbalization items was .15 and .18 for third- and sixth-grade children, respectively. Both correlations are nonsignificant. However, when composite scores were used, the mean correlations between reading ability and verbalization increased to a respectable level, .37 and .42 for third-and sixth-grade children, respectively.

In addition to the correlational analyses, we used the composite scores to classify children as mature metacognizers (high on both performance and verbalization), in transition (high performance, low verbalization), mimickers (low performance, high verbalization), and low, or immature, metacognizers (low on both performance and verbalization). In all cases (use of important information, memory, decoding, comprehension, and strategies), only the older/better readers tended to be classified as mature metacognizers. Younger/poorer readers usually were classified as low, or immature, metacognizers. On the whole, very few children were classified as in transition or as mimickers (except on the decoding skills). It appears, then, that a relation between knowledge and strategy use is present in reading, and that more optimism about cognitive–metacognitive connections is warranted than has been expressed in some summaries (Cavanaugh & Perlmutter, 1982; Cavanaugh, 1982). Also, Schneider (in press) provides a comprehensive and optimistic documentation of cognitive–metacognitive linkages in the area of memory.

The relation of monitoring to strategy use has received even less attention than the knowledge/use issue, and the knowledge/monitoring/use issue has been virtually ignored. However, we did report data that suggested that older/better readers not only are more apt to adjust their reading to meet the demands of various purposes (to skim, to study, etc.), but they also are more apt to monitor comprehension. In addition, older/better readers tended to know more about different strategies that could be used if comprehension breaks down. Much more research is needed on the nature of the interconnections between knowledge, monitoring, and strategy use, but initial indications are encouraging (see Chi, in press, for additional commentary on these issues). If successful strategy use during reading depends not only on the child's ability to use a particular strategy but also on his or her knowledge of and ability to monitor strategy use, then the problem of training mature study skills becomes very complex. This issue will be examined in depth in a following section.

At this point, we feel that it is reasonable to argue that since performance and verbalization measures covary to such a large degree, it is likely that the causal relationship between the two is an interactive one. As the child learns to use more skills, he or she is able to verbalize more about these skills. In addition, as the child acquires more knowledge about skills, there is more of a tendency to use these skills in a performance situation. In addition, it also should be noted that since performance and verbalization measures covary to such a large degree, it makes little difference whether the researcher uses the traditional definition of "meta" or the more restrictive one used in this investigation.

The next question that should be addressed is whether or not using a "meta" measure is a useful addition to any investigation. Certainly, the addition of the interview sessions added considerable time and expense to this study. What was gained, if anything, by this additional information? If the researcher is interested in predicting reading ability, the addition of the verbalization measures adds relatively little. Measures of performance consistently are the best predictors of reading ability. In fact, verbalization measures are relatively poor predictors of reading ability, particularly at the sixth-grade level. As has been mentioned, if verbalization is an important measure to consider, it appears that it would be in connection with language at the third-grade level and with monitoring comprehension at the sixth-grade level. However, in spite of these results, we do not advocate the conclusion that verbalization skills are not important, because high performance rarely is found in the absence of high verbalization scores. Of course, it is impossible to conclude that verbalization skills are *necessary* for high performance.

Yet in spite of the fact that the verbalization measures made a relatively poor showing in terms of predictive power, these measures did provide a great deal of valuable information. In many cases, the information took the form of qualitative rather than quantitative data about the children involved in the study. Using information from the interviews, we were able to make a fairly complete description of the two levels of reading ability and the knowledge that readers at each level have. In addition, the gradual yet continuous transition from beginning to mature reading is obvious from the interview data. Indeed, the interpretation of the performance variables was aided at times by information from the interviews. For example, the reluctance of older/better readers to skim a story for one specific piece of information was puzzling. However, information from the interviews suggested valid reasons for such performance. Therefore, if the children have reported accurately, there is little question as to how to interpret the performance findings.

If future "meta" research is to be fruitful, though, more attention needs to be given to the development of theory and methodology in this area. For example, as yet there has been little concern for the development of a theoretical framework that would explain how and why metacognition is important to learning. In addition, no one has examined the possibility that differences in items (i.e., the way in which questions are asked, the amount of probing done by the interviewer, etc.) may affect the results. In dealing with the issue of probing done by the interviewer, for example, some have suggested that it would be best to score only the child's first responses (e.g., Myers & Paris, 1978). However, in this study, a standard probe ("Anything else?") was used until the child indicated that he or she could think of nothing else, or until the child began to repeat things that had already been said. The entire response, then, was scored, thus giving the child every possible opportunity to express knowledge on any particular item. In cases where only the initial response is scored, the level of development of the child is apt to be underestimated. However, by using the present scoring system, qualitative differences—obvious to the interviewer and anyone reading the transcripts—were ignored. For example, a child who indicated one piece of knowledge after the initial question, another after the first probe, and a third after an additional probe often would be given the same score as a child who spontaneously produced three pieces of information after the initial question. In addition, the way in which the initial queston was asked, or the level of difficulty of the question, could make a difference in the nature of the response. For example, in some sections various questions were asked in order to assess the same type of knowledge (e.g., grammatical acceptability of sentences in the language section). A child usually responded at approximately the same level on all questions, but there was individual variance between items.

All in all, the whole question of "What is meta?" still is open to debate and obviously affects the scoring method that any particular researcher chooses to use.

Educational Implications

When faced with the extensive data generated by this study, there is a compulsion to try to relate the findings to educational practices. However, in view of the subtleties and complexities of the data and the issues involved, we feel that any implications should be stated with caution and regarded purely as speculations.

Indeed, many already have suggested that it may be beneficial to train metacognitive skills with the aim of increasing performance (Flavell, 1978; Meichenbaum & Asarnow, 1979; Myers & Paris, 1978). However, the problems involved in any training situation also have been noted. For example, Meichenbaum and Asarnow (1979) suggested that explicit objectives are important in training studies. They also stressed the difficulty of obtaining generalization of acquired skills to new situations. Flavell (1978) concluded that any training program that encouraged monitoring of cognitive processing should, itself, be subject to careful monitoring. In view of the fact that this study sheds little, if any, light on the causation issue, it is very difficult to suggest what should be trained or why. Indeed, the metacategorizations indicate that it would be possible to train mimickers who could not use the skills involved (e.g., decoding in sixth-grade poor readers). If performance is not improved by training, then the benefit of that training is highly questionable.

In addition, it appears that high reading ability is the result of a composite of abilities in each of the areas tested in this investigation. ("Deficit" readers, children who were low on one score or set of scores and high on all others, were found only occasionally.) As a result, it would be difficult to pinpoint explicit objectives, as suggested by Meichenbaum and Asarnow (1979) that would benefit from training. In view of these considerations, it is suggested strongly that it would not be wise to leap quickly from the present findings to practical applications. There are many questions to be considered before specific educational implications can be justified. In spite of this caution, it also should be noted that verbalization measures did aid in explaining many of the performance results and, as such, may be useful in a practical sense. The implication of this observation, put simply, is that teachers, perhaps, should talk to students about how and what they know and why they do things the way they do.

The Effect of Schooling and/or Age

Throughout all of the results sections, it is impossible not to notice that the grade/age effect usually was present and usually was at least as strong, if not stronger, than the effect of reading ability. It would be foolhardy to try to argue that schooling did not affect the maturity of both the use of skills and the manner of expressing one's knowledge about cognitive skills. However, there are notable exceptions to this grade/age difference. In particular, it often was found that third-grade good readers and sixth-grade poor readers were performing at the same level on many tasks (even when reading stories of the same level of difficulty, as shown in Appendix I). The difference between the two groups appears to be a qualitative rather than a quantitative one. Once again, support for this conclusion comes from the interviews. Indications were that the sixth-grade poor readers had acquired coping strategies rather than reading strategies, in that they knew of ways to get out of reading situations. In addition, unlike the sixth-grade good readers, poor readers at this level were perfectly willing to use a skim strategy. On the other hand, third-grade good readers often indicated that they enjoyed reading and read a great deal at home. One third-grade good reader repeatedly asked if he could have copies of the experimental stories so that he could reread them at home.

Conceptions of Reading

How does the information gained from this investigation affect our concept of reading? In light of the results, we cannot resist the temptation to speculate about the nature of the development of reading processes.

As outlined in the Introduction, conceptions of reading as involving knowledge, monitoring, and strategy use are not entirely new. In fact, many early reading theorists, including Huey (1908/1968) and Thorndike (1917), acknowledged the importance of monitoring and other study skills to aid in reading comprehension. For example, Huey (1908/1968) included note taking, referencing, and index usage as aids to comprehension. In more recent years, reading has been considered as a complex problem-solving situation (e.g., Bransford, Stein, Shelton & Owings, 1980; Kachuck & Marcus, 1976; Kavale & Schreiner, 1979; Olshavsky, 1977; Reid, 1966). Also, there are excellent review and position papers on the metacognitive research that has been conducted (e.g., Baker & Brown, in press; Brown, 1978, 1980; Brown & Deloache, 1978; Flavell & Wellman, 1977; Paris, 1978; Yussen, Matthews & Hiebert, 1982).

Within the general context, we think it not unreasonable to view reading within much the same framework as the model of cognitive processing proposed by Miller, Galanter, and Pribram (1960). This model was outlined in Chapter 1) and basically consists of executive or control processes (metaplans) that guide behavior (through TOTE units) and generate cognitive actions (Plans).

When beginning to read, the child first must learn some means of rapidly recognizing single words. There are, of course, several cognitive strategies ("plans" in the above-discussed vernacular) that can be used in order to decode words, such as recognition of features of the whole word, "sounding out" the word (e.g., Chall, 1967), asking someone what the word says, or guessing the word from the context or from a single feature (such as an initial letter). The reader also could decide to skip the word completely, thus avoiding the immediate problem. Knowledge about each of these strategies might affect the way in which the child approaches a reading task. Indeed, most third-grade readers know about different decoding strategies for single words. Moreover, while decoding, the child must monitor progress. This involves knowing whether or not any particular word has been dealt with successfully. If the reader realizes that decoding has failed (i.e., he or she is unable to pronounce the word), then knowledge about various decoding strategies should affect subsequent action. If a child knows about several possible strategies of decoding, this increases the likelihood that the word will be decoded. When more is known about each particular decoding strategy, the probability of choosing the most efficient strategy, given the situation, increases. Indeed, older/better readers are more likely both to *know about* and to *use effectively* a variety of decoding skills. That is, they possess both plans and metaplans that younger readers do not.

Even when words are decoded to sound, it is possible that appropriate meanings will not be constructed or accessed. Once again, the reader must be able to monitor performance by deciding whether or not what has been read has been understood (e.g., Adams, 1980). Only older/better readers know that there is a difference between what a word "says" and what a word "means." If decoding has failed at this stage, then the reader could ask someone what the word means or could look the word up in a dictionary. Knowledge of these strategies, and the effectiveness of each, should affect subsequent action.

Reading, with respect to decoding, is concerned not only with words, but also with larger units such as sentences. Certainly, the reader must be able to monitor performance on these larger units. If comprehension fails, for example, at the sentence level, the reader can figure out what the sentence says by using contextual cues, asking someone else, or looking up words in a dictionary. The reader also could decide to skip the sentence completely, thus avoiding the immediate problem. Knowledge of each of these strategies, and the efficiency of each, should affect subsequent action. Only older/better readers can provide sophisticated information about how to decode sentences. Consistent with the earlier general comments, reading in a decoding sense involves not only the use of cognitive strategies, but also knowledge of these strategies and the active monitoring of performance.

In addition to decoding strategies, mature readers have available a number of different reading strategies to meet the demands of various situations. For example, they may reread, skim read, paraphrase, or concentrate on important units. Only the older/better reader can express knowledge of many of these strategies (also see Kobasigawa, Ransom & Holland, 1980; Myers & Paris, 1978). The knowledge that the reader has about these cognitive strategies may affect how he or she approaches the reading task. For example, the reader should be able to evaluate task demands and choose the most efficient method of reading for any particular situation. While reading, a child might need to be able to identify the important units of text in order to carry out an appropriate reading strategy (e.g., Brown & Smiley, 1977). In addition, the child needs to monitor progress. The reader must realize when important information is read, decide whether or not the purpose of reading is being achieved, and determine if a change in strategy is needed. Once again, research indicates that older/better readers are more proficient at monitoring comprehension (also see Flavell, Speer, Green & August, 1981; Markman & Gorin, 1981; Baker, 1979; Forrest-Pressley, 1984; Winograd & Johnson, 1980). If the reader recognizes comprehension failure, it follows that the reader's knowledge about various reading strategies and the efficency of each should affect subsequent behavior. The reader might realize that the problem is a matter of decoding and, thus, decide to change decoding strategies. However, it also is possible that the reader might realize that important information is not being retained and, thus, decide that a change in advanced reading strategy is necessary.

In our view, data such as our own and the other studies cited above provide some validation for three major components of flexible strategy usage in the mature reader: (1) knowledge of possible alternative strategies, (2) spontaneous use of reading strategies, and (3) active monitoring and adjustment of strategies. As can be noted from the above arguments, we do not feel that it is sufficient for the child to be able to use the cognitive strategies connected with reading. Rather, an important aspect involves the active monitoring of the cognitive processes (e.g., comprehension) and the use of knowledge about reading strategies to predict efficiency and plan future behavior. The metacognitive aspect of reading, then, involves the control of appropriate cognitive skills, in the sense of planning cognitive activities, choosing among alternative activities, monitoring the performance of activities, and changing activities.

As we have suggested, during the period in which reading is being taught, the child is developing cognitive and metacognitive competencies in the areas of language, attention, and memory. Unfortunately, the exact relationship between each component process and each reading skill is difficult to assess. Most variables are intercorrelated, and we have little idea of the stability or precise nature of the relationships. If "deficits" had been found in the developmental processes that matched "deficits" in reading skills, then the relationship question would have been much easier to answer. However, "deficit" readers were found only rarely.

Much of the above conception of reading obviously is speculative to a large degree. However, the consistent pattern of results found in this investigation does lead to the conclusion that the reading process has both cognitive and metacognitive aspects. Metacognition, or "executive functions," direct the flexible use of various strategies in our world view. While the presented conception of reading skills is, of course, speculative, it directs attention to areas in which future research should be concentrated. For example, does it follow that an increased knowledge base automatically increases the child's chance of choosing the most efficient strategy for any particular situation? Is it necessary that the knowledge base be explicit? Does the knowledge base have any direct effect on the monitoring system, or is the monitoring system an independent process? If the monitoring system identifies an error, what conditions increase the probability that the error will be corrected? What conditions lead to an error being undetected or uncorrected, and what are the consequences of such failures? What conditions produce a monitoring individual as opposed to a nonmonitoring individual? Is it possible to train complex metacognitive skills?

In this chapter, we have taken the position that metacognition is a necessary component of flexible strategy use during reading. However, we urgently need data collected in controlled situations that allow inferences about the directions of the effects. Indeed, we would like to stress that we do not believe that simply training knowledge or "meta" will be the answer, since it appears that it is quite possible to create mimickers (i.e., children who can mimic a verbal response but do not use the knowledge to increase performance). Rather, we expect that training will be a complex procedure, perhaps involving the introduction of specific knowledge, practice with using different strategies, evaluation of the effectiveness of various strategies, practice at monitoring, comparison of good and poor messages, modeling of approriate responses, and feedback on predictions of success. Research involving each of these components in a practical setting is needed. The tremendous gap between research and practice is appalling, as is the lack of empirical evaluation of such things as study techniques, which obviously overlap with our major concerns. Indeed, instructional research must be conducted on these problems, and it is hoped that the research will be conducted in such a fashion so as to add to our knowledge of the actual processes involved (Belmont & Butterfield, 1977).

In spite of an obvious lack of progress in developing study-skills programs that can be substantiated by empirical evidence, an historical overview does point to an interesting recurring notion. This notion is that at all costs, the reader must be motivated to become an "active" learner (i.e., "keeping the mind active," Swain, 1917). This is attempted by suggesting that the reader do certain things that will insure an active involvement with the material (e.g., question, recite, review, take notes, underine, summarize, visualize, monitor progress). Certainly, one problem inherent in this notion is that the child must not only know how to use these specific skills, but also must know what information should be questioned, recited, reviewed, underlined, and visualized. In addition, a child must continually monitor performance and check progress against the desired goal. These skills definitely are metacognitive in nature, yet traditional study-skills programs have not emphasized the metacognitive aspect of reading to learn. It is only in recent years that researchers and practitioners have begun to consider knowledge and monitoring as essential to reading for meaning (e.g., Baker, in press; Baker & Brown, in press; Yussen, Matthews & Hiebert, 1982). We hope that, within the next several years, researchers and practitioners will begin to bridge the gap between research and practice, and develop empirically substantiated methods of teaching study skills. In the process, we might also come to know more about cognition, metacognition, the connections between them, and what it all has to do with being a skilled reader.

References

Adams, M. J. Failures to comprehend and levels of processing in reading. In R. J. Spiro, B. C. Bruce, & W. F. Brewer (Eds.), *Theoretical issues in reading comprehension*. Hillsdale, NJ: Erlbaum, 1980.

Adams, M., & Bruce, B. Background knowledge and reading comprehension. Reading Education Report No. 13, Center for the Study of Reading, University of Illinois, Urbana-Champaign, 1980.

Allen, D. Some effects of advance organizers and level of question on the learning and retention of written social studies material. *Journal of Educational Psychology*, 1970, *61*, 333-339.

Alta-Boyd Test of Phonic Skills. Unpublished test, unknown author, University of Alberta.

Anderson, R. C., & Biddle, W. B. On asking people questions about what they are reading. In G. Bower (Ed.), *The psychology of learning and motivation* (Vol. 9). New York, NY: Academic Press, 1975.

Anderson, T. H. Study strategies and adjunct aids. In R. J. Spiro, B. C. Bruce, & W. F. Brewer (Eds.), *Theoretical issues in reading comprehension*. Hillsdale, NJ: Erlbaum, 1980.

Anderson, T. H., & Armbruster, B. B. Studying. Technical Report No. 155, Center for the Study of Reading, University of Illinois, Urbana-Champaign, 1980.

Ausubel, D. In defense of advance organizers: A reply to the critics. *Review of Educational Research*, 1978, *48*, 251-257.

Baker, L. Comprehension monitoring: Identifying and coping with text confusions. Technical Report No. 145, Center for the Study of Reading, University of Illinois, Urbana-Champaign, 1979.

Baker, L. How do we know we don't understand? Standards for evaluating text comprehension. In D. Forrest-Pressley, G. E. MacKinnon, & T. G. Waller (Eds.), *Cognition, metacognition and human performance*. New York, NY: Academic Press, in press.

Baker, L., & Anderson, R. L. Effects of inconsistent information on text processing: Evidence for comprehension monitoring. *Reading Research Quarterly*, 1982, *17*, 281-294.

Baker, L., & Brown, A. L. Metacognitive skills in reading. In D. Pearson, M. Kamil, R. Barr, & P. Mosenthal (Eds.), *Handbook of reading research.* New York, NY: Longman, in press.

Barnes, B., & Clawson, E. Do advance organizers facilitate learning? Recommendations for further research based on an analysis of 32 studies. *Review of Educational Research,* 1975, *45,* 637-659.

Barr, R. The effect of instruction on pupil reading strategies. *Reading Research Quarterly,* 1975, *10,* 555-582.

Barron, R. W. Development of visual word recognition: A review. In G. E. MacKinnon & T. G. Waller (Eds.), *Reading research: Advances in theory and practices* (Vol. 2). New York, NY: Academic Press, 1981.

Belmont, J. M., & Butterfield, E. D. The instructional approach to developmental cognitive research. In R. Kail & J. Hagen (Eds.), *Perspectives on the development of memory and cognition.* Hillsdale, NJ: Erlbaum, 1977.

Bialystok, E., & Ryan, E. A metacognitive framework for the development of first and second language skills. In D. Forrest-Pressley, G. E. MacKinnon, & T. G. Waller (Eds.), *Cognition, metacognition and human performance.* New York, NY: Academic Press, in press.

Bloom, L., & Lahey, M. *Language development and language disorders.* New York, NY: Wiley, 1978.

Borkowski, J. G., Reid, M., & Kurtz, B. Metacognition and retardation: Paradigmatic, theoretical and applied perspectives. In R. Sperber, C. McCauley, & P. Brooks (Eds.), *Learning and cognition in the mentally retarded.* Baltimore, MD: University Park Press, in press.

Bransford, J., Stein, B., Shelton, T., & Owings, R. Cognition and adaption: The importance of learning to learn. In J. Harvey (Ed.), *Cognition, social behavior and the environment.* Hillsdale, NJ: Erlbaum, 1980.

Brown, A. Metacognitive development and reading. In R.J. Spiro, B. Bruce & W. F. Brewer (Eds.), *Theoretical issues in reading comprehension.* Hillsdale, N.J.: Erlbaum, 1980.

Brown, A., & DeLoache, J. Skills, plans and self-regulation. In R. S. Siegler (Ed.), *Children's thinking: What develops?* Hillsdale, NJ: Erlbaum, 1978.

Brown, A., & Smiley, S. Rating the importance of structural units of prose passages: A problem of metacognitive development. *Child Development,* 1977, *48,* 1-8.

Brown, A. L. Knowing when, where, and how to remember: A problem of metacognition. In R. Glaser (Ed.), *Advances in instructional psychology* (Vol. 1). Hillsdale, NJ: Erlbaum, 1978.

Brown, A. L., Campione, J. C., & Day, J. Learning to learn: On training students to learn from texts. *Educational Researcher,* 1981, *10(2),* 14-21.

Carroll, J. B. Defining language comprehension: Some speculations. In J. B. Carroll & R. O. Freedle (Eds.), *Language comprehension and the acquisition of knowledge.* New York, NY: Winston, 1972.

Cavanaugh, J. C. Metamemory-strategy relationships: A new chapter for Bulfinch or our Rosetta stone? Paper presented at the annual meeting of the American Educational Research Association, New York, NY: March 1982.

Cavanaugh, J. C., & Borkowski, J. G. Searching for metamemory-memory connections: A developmental study. *Developmental Psychology,* 1980, *16,* 441-453.

Cavanaugh, J. C., & Perlmutter, M. Metamemory: A critical examination. *Child Development,* 1982, *53,* 11-28.

Chall, J. *Learning to read: The great debate*. New York, NY: McGraw-Hill, 1967.

Chi, M. Interactive roles of knowledge and strategies in development. In S. Chapman, J. Segal, & R. Glaser (Eds.), *Thinking and learning skills: Current research and open questions* (Vol. 2). Hillsdale, NJ: Erlbaum, in press.

Clawson, E., & Barnes, B. The effects of organizers on the learning of structured anthropology materials in the elementary grades. *Journal of Experimental Education,* 1973, 42, 11-15.

Dale, P. *Language development: Structure and function*. Hillsdale, IL: Dryden Press, 1972.

Dennis, M. Unpublished test of grammatical and syntactic acceptability. Hospital for Sick Children, Toronto, Ontario, Canada.

deVilliers, J., & deVilliers, P. *Language acquisition*. Cambridge, MA.: Harvard University Press, 1978.

Downing, J., & Oliver, P. The child's conception of 'a word.' *Reading Research Quarterly,* 1974, *9,* 568-582.

Ehri, L. Word consciousness in readers and prereaders. *Journal of Educational Psychology,* 1975, *67,* 204-212.

Ehri, L. Word learning in beginning readers and prereaders: Effects of form class and defining contexts. *Journal of Educational Psychology,* 1976, *68(b),* 832-842.

Faw, H., & Waller, T. G. Mathemagenic behaviors and efficiency in learning from prose materials: Review, critique and recommendations. *Review of Educational Research,* 1977, *46,* 391-420.

Flavell, J. *Cognitive development*. Englewood Cliffs, NJ: Prentice-Hall, 1977.

Flavell, J. Metacognition. Paper presented in a symposium on *Current perspectives on awareness and cognitive processes* at the meeting of the American Psychological Association, Toronto, Ontario, Canada, 1978.

Flavell, J. Metacognitive aspects of problem solving. In L. B. Resnick (Ed.), *The nature of intelligence*. Hillsdale, NJ: Erlbaum, 1976.

Flavell, J., Speer, J. R., Green, F. L., & August, D. L. The development of comprehension monitoring and knowledge about communication. *Monographs of the Society for Research in Child Development,* 1981, *46* (No. 5, Serial No. 192).

Flavell, J., & Wellman, H. Metamemory. In R. Kail & J. Hagen (Eds.), *Perspectives on the development of memory and cognition*. Hillsdale, NJ: Erlbaum, 1977.

Ford, N. Recent approaches to the study and teaching of "effective learning" in higher education. *Review of Educational Research,* 1981, 51, 345-377.

Forrest, D. L., & Barron, R. Metacognitive aspects of the development of reading skill. Paper presented at the biennial meeting of the Society for Research in Child Development, New Orleans, LA, 1977.

Forrest-Pressley, D. L. Comprehension monitoring in reading. Unpublished manuscript, Children's Psychiatric Research Institute, London, Ontario, 1984.

Forrest-Pressley, D. L., & Gillies, L. A. Children's flexible use of strategies during reading. In M. Pressley & J. R. Levin (Eds.), *Cognitive strategy research: Educational applications*. New York, NY: Springer-Verlag, 1983.

Frase, L. T. Maintenance and control in the acquisition of knowledge from written materials. In J. B. Carroll & R. O. Freedle (Eds.), *Language comprehension and the acquisition of knowledge*. New York, NY: Winston, 1972.

Frase, L. T. Prose processing. In G. Bower (Ed.), *The psychology of learning and motivation* (Vol. 9). New York, NY: Academic Press, 1975.

Frase, L. T. Purpose in reading. In J. Guthrie (Ed.), *Cognition, curriculum and comprehension*. Newark, NJ: International Reading Association, 1977.

Furth, H. G. Reading as thinking: A developmental perspective. In F. B. Murray & J. J. Pikulski (Eds.), *The acquisition of reading: Cognitive, linguistic and perceptual prerequisites*. Baltimore, MD: University Park Press, 1978.

Gates, A., & MacGinitie, W. *Gates-MacGinitie Reading Tests, Primary C: Form 1 & Survey D: Form 1*. Teachers College, Columbia University, New York, NY: Teachers College Press, 1972.

Gibson, E. Reading for some purpose. In J. Kavanagh & I. Mattingly (Eds.), *Language by ear and by eye*. Cambridge, MA: The MIT Press, 1972.

Gibson, E., & Levin, H. *The psychology of reading*. Cambridge, MA: The MIT Press, 1975.

Golinkoff, R. A comparison of reading comprehension processes in good and poor comprehenders. *Reading Research Quarterly,* 1976, *4,* 621-659.

Goodman, K. Analyses of oral reading miscues. *Reading Research Quarterly,* 1969, *5,* 9-30.

Grueneich, R., & Trabasso, T. The story as social environment: Children's comprehension and evaluation of intentions and consequences. Technical Report No. 142, Center for the Study of Reading, University of Illinois, Urbana-Champaign, 1979.

Gunning, R. *The technique of clear writing*. New York, NY: McGraw-Hill, 1968.

Huey, E. B. *The psychology and pedagogy of reading*. Cambridge, MA: The MIT Press, 1968. (Originally published by MacMillan, 1908.)

Kachuck, B., & Marcus, A. Thinking strategies and reading. *Reading Teacher,* 1976, *30,* 157-161.

Kavale, K., & Schreiner, R. The reading processes of above average and average readers: A comparison of the use of reasoning strategies in responding to standardized comprehension measures. *Reading Research Quarterly,* 1979, *15,* 102-128.

Kavanagh, J. F., & Mattingly, I. G. (Eds.), *Language by ear and by eye*. Cambridge, MA: The MIT Press, 1972.

Kobasigawa, A., Ransom, C., & Holland, C. Children's knowledge about skimming. *Alberta Journal of Educational Research,* 1980, *26,* 169-181.

Kreutzer, M., Leonard, C., & Flavell, J. An interview study of children's knowledge about memory. *Monographs of the Society for Research in , Child Development,* 1975, *40,* (No. 1, Serial No. 159).

LaBerge, D., & Samuels, J. Toward a theory of automatic information processing in reading. *Cognitive Psychology,* 1974, *6,* 293-323.

Levin, J. R., & Pressley, M. Improving children's prose comprehension: Selected strategies that seem to succeed. In C. Santa & B. Hayes (Eds.), *Children's prose comprehension: Research and practice*. Newark, DE: International Reading Association, 1981.

Lorge, I., Thorndike, R., Hagen, E., & Wright, E. *Canadian Lorge-Thorndike Intelligence Tests, Levels A & D*. Toronto: Thomas Nelsen & Sons (Canada) Ltd., 1967.

Mackworth, J. Some models of the reading process: Learners and skilled readers. *Reading Research Quarterly,* 1972, *7,* 701-733.

Markman, E. Realizing that you don't understand: A preliminary investigation. *Child Development,* 1977, *48,* 986-992.

Markman, E. Realizing that you don't understand: Elementary school children's awareness of inconsistencies. *Child Development,* 1979, *50,* 643-655.

Markman, E., & Gorin, L. Children's ability to adjust their standards for evaluating comprehension. *Journal of Educational Psychology,* 1981, *73,* 320-325.

Meichenbaum, D., & Asarnow, J. Cognitive-behaviour modification and meta-cognitive development: Implications for the classroom. In P. C. Kendall and S. D. Hollon (Eds.), *Cognitive-behavioral interventions: Theory, research and procedures.* New York, NY: Academic Press, 1979.

Miller, G., Galanter, E., & Pribram, K. *Plans and the structure of behavior.* New York, NY: Holt, 1960.

Miller, P. Metacognition and attention. In D. Forrest-Pressley, G. E. MacKinnon, & T. G. Waller (Eds.), *Cognition, metacognition and human performance.* New York, NY: Academic Press, in press.

Murray, F. B., & Pikulski, J. (Eds.), *The acquisition of reading.* Baltimore, MD: University Park Press, 1978.

Myers, M., & Paris, S. Children's metacognitive knowledge about reading. *Journal of Educational Psychology,* 1978, *70,* 680-690.

Nisbett, R. E., & Wilson, T. D. Telling more than we can know: Verbal reports on mental processes. *Psychological Review,* 1977, *84,* 231-259.

Olshavsky, J. Reading as problem solving: An investigation of strategies. *Reading Research Quarterly,* 1976-77, *12(4),* 654-674.

O'Sullivan, J., & Pressley, M. Completeness of strategy instruction. Unpublished manuscript, University of Western Ontario, 1983.

Otto, W., & White, S. (Eds.), *Reading expository text.* New York, NY: Academic Press, 1982.

Paris, S. Metacognitive development: Children's regulation of problem-solving skills. Paper presented at the Midwestern Psychological Association, Chicago, IL, May 1978.

Pichert, J. Sensitivity to what is important in prose. Technical Report No. 149, Center for the Study of Reading, University of Illinois, Urbana-Champaign, 1979.

Pressley, M., Borkowski, J. G., & O'Sullivan, J. Children's metamemory and the teaching of memory strategies. In D. Forrest-Pressley, G. E. MacKinnon, & T. G. Waller (Eds.), *Cognition, metacognition and human performance.* New York, NY: Academic Press, in press.

Pressley, M., Heisel, B., McCormick, C., & Nakamura, G. Memory strategy instruction with children. In C. J. Brainerd & M. Pressley (Eds.), *Verbal processes in children.* New York, NY: Springer-Verlag, 1982.

Proger, B., Carter, C., Mann, L., Taylor, R., Bayuk, R., Morris, V., & Reckless, D. Advance and concurrent organizers for detailed verbal passages used with elementary school pupils. *Journal of Educational Research,* 1973, *66,* 451-456.

Reder, L. Comprehension and retention of prose: A literature review. Technical Report No. 108, Center for the Study of Reading, University of Illinois, Urbana-Champaign, 1978.

Reid, J. Learning to think about reading. *Educational Research,* 1966, *9,* 56-62.

Rothkopf, E. Structural text features and the control of processes in learning from written materials. In J. B. Carroll & R. O. Freedle (Eds.), *Language comprehension and the acquisition of knowledge*. New York, NY: Winston, 1972.

Rothkopf, E., & Billington, M. A two-factor model of the effect of goal-descriptive directions on learning from text. *Journal of Educational Psychology,* 1975, 67, 692-704.

Royer, J., Hastings, C. N., & Hook, C. A sentence verification technique for measuring reading comprehension. Technical Report No. 137, Center for the Study of Reading, University of Illinois, Urbana-Champaign, 1979.

Rushton, V. P., Brainerd, C. J., & Pressley, M. Behavioral development and construct validity: The principle of aggregation. *Psychological Bulletin,* 1983, *94,* 18-38.

Ryan, E., McNamara, S., & Kenny, M. Linguistic awareness and reading ability among beginning readers. *Journal of Reading Behavior,* in press.

Schneider, W. Developmental trends in the metamemory-memory behavior relationship: An integrative review. In D. Forrest-Pressley, G. E. MacKinnon, & T. G. Waller (Eds.), *Cognition, metacognition and human performance*. New York, NY: Academic Press, in press.

Singer, H., & Ruddell, R. (Eds.), *Theoretical models and processes of reading*. Newark, NJ: International Reading Association, 1976.

Slosson, R. *Slosson Oral Reading Test (SORT)*. East Aurora, NY: Slosson Educational Publishing Inc., 1963.

Smith, F. *Understanding reading*. New York, NY: Holt, Rinehart, & Winston, 1971.

Spiro, R. J., Bruce, B., & Brewer, W. F. (Eds.), *Theoretical issues in reading comprehension*. Hillsdale: NJ: Erlbaum, 1980.

Stanovich, K., & West, R. Mechanisms of sentence context effects in reading: Automatic activation and conscious attention. *Memory and Cognition,* 1979, *7,* 77-85.

Sternberg, R. J. The development of human intelligence. Technical Report No. 4, Cognitive Development Series, Department of Psychology, Yale University, New Haven, CT, 1979.

Swain, G. F. *How to study*. New York, NY: McGraw-Hill, 1917.

Swensen, I., & Kulhavy, R. Adjunct questions and the comprehension of prose by children. *Journal of Educational Psychology,* 1974, *66,* 212-215.

Thorndike, E. Reading as reasoning: A study of mistakes in paragraph reading. *Journal of Educational Psychology,* 1917, *8,* 323-330.

Thorndyke, P. W. Cognitive structures in comprehension and memory of narrative discourse. *Cognitive Psychology,* 1977, *9,* 77-110.

Tierney, R. J., & Cunningham, J.W. Research on teaching reading comprehension. Technical Report No. 187, Center for the Study of Reading, University of Illinois, Urbana-Champaign, 1980.

Trabasso, T. On the making of inferences during reading and their assessment. Technical Report No. 157, Center for the Study of Reading, University of Illinois, Urbana-Champaign, 1980.

Walker, C., & Meyer, B. Integrating information from text: An evaluation of current theories. *Review of Educational Research,* 1980, *50,* 421-437.

Waterhouse, L., Fischer, K., & Ryan, E. *Language awareness and reading*. Newark, DE: International Reading Association, 1980.

Willows, D. Reading between the lines: Selective attention in good and poor readers. *Child Development,* 1974, *45,* 408-415.

Winograd, P., & Johnson, P. Comprehension monitoring and the error detection paradigm. Technical Report No. 153, Center for the Study of Reading, University of Illinois, Urbana-Champaign, 1980.

Wong, B., & Jones, W. Increasing metacomprehension in learning-disabled and normally-achieving students through self-questioning techniques. *Learning Disability Quarterly,* 1982, *5,* 228-240.

Wong, B. Metacognition and learning disabilities. In D. Forrest-Pressley, G. E. MacKinnon, & T. G. Waller (Eds.), *Cognition, metacognition and human performance.* New York, NY: Academic Press, in press.

Yussen, S. R., & Berman, L. Memory predictions for recall and recognition in first-, third- and fifth-grade children. *Developmental Psychology,* 1981, *17,* 224-229.

Yussen, S. R., Matthews, S., & Hiebert, E. H. Metacognitive aspects of reading. In W. Otto & S. White (Eds.), *Reading expository test.* New York, NY: Academic Press, 1982.

APPENDIX A

List of Verbalization Items (Abstracted) with Interrater Reliability (Percent Agreement)

Code	Rel.	Item
Decoding Verbalization		
D-V-1	89	What do you do when you come to a word that you do not know?
D-V-2	90	Is there a difference between knowing what a word "says" and knowing what a word "means"?
D-V-3	88	Is it better to sound out a word that you do not know or ask someone what it says?
D-V-4	83	What do you do when you come to a whole sentence that you do not understand?
Comprehension Verbalization		
C-V-1	81	How do you know when you are ready to write a test?
C-V-2	81	Would you know how well you had done on the test before you got it back? How?
C-V-3	86	How would you know when you knew enough about a game to be able to teach someone else about it?
Strategies Verbalization		
S-V-1	83	What do you do when you read in preparation for a test?
S-V-2	97	Is there anything that you can do to make what you are reading easier to remember?
S-V-3	85	How would you find the name of a place in a story?

S-V-4	82	How would you remember a story so that you could tell it to a friend later?
S-V-5	90	How much of the story would you remember?
S-V-6	85	How would you think of a title for a story?

Language Verbalization

L-V-1	85	What can you tell me about words?
L-V-2	85	Where do we use words?
L-V-3	72	What can you tell me about sentences?
L-V-4	74	Where do we use sentences?
L-V-5	86	Is "tdet" a word/not a word? Why?
L-V-6	89	Is "meff" a word/not a word? Why?
L-V-7	83	Is "stone" a word/not a word? Why?
L-V-8	90	Is there anything you could do if you were not sure?
L-V-9	90	(If more than one was judged to be a word.) Which one is the best word?
L-V-10	81	John park to went.
L-V-11	93	Jane played with her friends.
L-V-12	97	After school, Bill wented home.
L-V-13	93	Before Mary could enter the contest.
L-V-14	92	My favorite dessert is radios with cream.
L-V-15	90	My favorite TV program are Gunsmoke.
L-V-16	93	My favorite toothpaste is Crest.
L-V-17	83	I paid the money by the man.
L-V-18	92	I paid the cash to the girl.
L-V-19	99	My favorite breakfast is eggs with bacon.
L-V-20	90	Can you always make things sound different? (Asked after transformations tasks.)
L-V-21	90	Can you always think of two meanings for every word? (Asked after homonyms task.)

Attention Verbalization

A-V-1	68	How would you decide which book you wanted to read from a library?
A-V-2	86	How would you decide who was a good reader?
A-V-3	88	What is the best way to decide if someone is a good reader?
A-V-4	89	How would you put words in alphabetical order?
A-V-5	82	How would you put rhyming words together?
A-V-6	89	How did you find so many so quickly? (Asked after redundancy-search task.)

A-V-7	94	How did you find so many so quickly? (Asked after nonredundancy-search task.)
A-V-8	94	Which search task was easier? Why?

Memory Verbalization

M-V-1	90	Would you get a drink of water or call right away after hearing a telephone number?
M-V-2	89	How do you remember a telephone number?
M-V-3	94	Would it be easier to remember these pictures with or without a story? Why?
M-V-4	94	How would you remember these pictures? (Clusterable.)
M-V-5	83	Anything else that you could do to remember these pictures?
M-V-6	71	How would you find a lost jacket?
M-V-7	79	How would you think of all possible ways to find a lost jacket?
M-V-8	93	Which friend would remember most of the names of the children at the party, the one who went straight home or the one who went to a play practice, where he met more people? Why?

APPENDIX B

Split-Half Reliability for Regression Analyses

F Values for Regression Equations

Grade 3		Grade 6	
Variable	*F*	Variable	*F*
Decoding			
P	62.60	P	64.36
IQ	4.03	IQ	18.19
V	4.01	V	0.74
Comprehension			
P	8.09	P	14.75
IQ	1.63	IQ	21.80
V	0.09	V	0.13
Strategies			
P	45.01	P	69.24
V	0.37	IQ	11.79
IQ	0.08	V	0.35
Reading			
D-P	19.58	S-P	29.55
S-P	19.15	C-P	14.36
C-P	16.30	D-P	12.75
C-V	1.64	IQ	5.21

S-V	0.12	D-V	1.42
D-V	0.43	C-V	0.12
IQ	0.27	S-V	0.05
Language			
P	15.96	P	30.89
V	8.07	IQ	13.16
IQ	1.43	V	0.25
Attention			
P	14.88	IQ	16.00
V	4.71	P	3.71
IQ	1.76	V	3.69
Memory			
P	10.32	IQ	9.46
IQ	6.03	P	4.44
V	0.11	V	3.73
Developmental Factors			
L-P	6.29	L-P	24.22
L-V	7.04	IQ	4.33
M-P	4.31	A-V	2.11
A-P	0.72	M-V	1.04
M-V	0.63	M-P	0.53
IQ	0.33	A-P	0.52
A-V	0.10	L-V	0.02
Complete			
D-P	9.62	S-P	23.52
L-V	2.59	C-P	9.99
S-P	14.83	L-P	5.09
C-P	14.04	D-P	4.94
C-V	2.41	IQ	1.51
M-P	2.50	D-V	3.17
M-V	1.13	A-P	1.42
L-P	0.09	C-V	0.89
IQ	0.11	A-V	0.15
S-V	0.08	M-V	0.21
A-V	0.03	S-V	0.14
D-P	0.03	L-V	0.09

A-P	*	M-P	*
Performance			
D-P	10.25	S-P	25.52
S-P	17.66	C-P	13.23
C-P	13.21	L-P	5.01
M-P	0.74	D-P	6.10
A-P	0.39	IQ	3.28
L-P	0.15	A-P	1.03
IQ	0.04	M-P	*
Verbalization			
L-V	10.50	IQ	13.79
C-V	7.52	C-V	12.68
A-V	1.67	L-V	4.66
IQ	1.66	A-V	2.86
S-V	1.02	S-V	1.18
M-V	0.50	D-V	0.13
D-V	0.03	M-V	*

Note: IQ refers to a nonverbal measure of intelligence.
* indicates that the variable did not enter the equation.

APPENDIX C

Decoding: Performance Items

D-P-1: Slosson Oral Reading Test. (Slosson, 1963.) The children were required to read out loud graded lists of words. One point was given for each word correctly pronounced. The total score was divided by two to obtain a maximum score of 100.

D-P-2: Nonsense Words. The children were required to read out loud the following list of words (from the Alta-Boyd Phonics Test). One point was given for each word correctly pronounced for a maximum total of 30. The "words" were as follows: clup, blam, gris, chas, shan, whes, thob, gam, cil, vill, kound, dight, tation, knet, phan, wrat, rafe, tife, sem, nid, rab, doil, moy, keat, sart, lirt, forn, lundle, vifted, delrim.

APPENDIX D

Decoding: Statistical Analyses

Table D-1. Decoding Performance Items: Analyses

Overall multivariate analysis

Grade, $F(3, 136) = 158.92$, $p < .0001$
Reading Ability, $F(6, 270) = 38.09$, $p < .0001$
Grade X Reading Ability, $p > .05$

Univariate F's

Item	Grade $df = (1, 138)$	Reading $df(2, 138)$	Grade X Reading $df(2, 138)$
D-P-1	442.03*	104.69*	0.49
D-P-2	88.38*	49.14*	0.21

*$p < .0001$

Table D-2. Decoding Performance Items: Correlations with Reading Ability

	Grade 3		Grade 6	
Item	r	partial r	r	partial r
D-P-1	.8012*	.7675*	.7874*	.7331*
D-P-2	.6435*	.5935*	.6464*	.6060*

*$p < .001$

Table D-3. Decoding Verbalization Items: Chi Square Results

Item	Effect of		
	Grade	Reading Ability	Other Significant
D-V-1	$\chi^2(1) = 13.78, p<.001$	n.s.	G.3 good vs. G.6 poor $\chi^2(1) = 4.94, p<.05$
D-V-2	$\chi^2(1) = 18.81, p<.0001$	$\chi^2(2) = 7.65, p<.05$	n.s.
D-V-3	n.s.	n.s.	n.s.
D-V-4	$\chi^2(1) = 7.36, p<.01$	n.s.	G.6 aver. vs. Gr.6 poor $\chi^2(1) = 4.41, p<.05$

Table D-4. Decoding Verbalization Items: Correlations with Reading Abilty

	Grade 3		Grade 6	
Item	r	partial r	r	partial r
D-V-1	.0980	.1117	−.0057	.0470
D-V-2	.2505*	.2881**	.3284**	.2495*
D-V-3	.1294	.1395	−.1060	−.1849
D-V-4	.0179	.0006	.2636*	.1910*

*$p < .05$ **$p < .01$

Table D-5. Decoding: Analysis of Computed Scores

Performance

Grade, $F(1, 138) = 231.86, p<.001$
Reading Ability, $F(2, 138) = 8.52, p<.001$
Grade X Reading Ability, $p>.05$

Verbalization

Grade, $F(1, 138) = 39.05, p<.001$
Reading Ability, $p>.05$
Grade X Reading Ability, $p>.05$

Table D-6. Decoding: Meta Categorizations

Overall, $\chi^2(15) = 92.27$, $p < .0001$
Grade, $\chi^2(1) = 40.94$, $p < .001$
Reading Ability, $\chi^2(2) = 8.40$, $p < .05$

APPENDIX E

Decoding: Verbalization Items

Question D-V-1: What do you do when you are reading and you come to a word that you don't know?

Scoring:

0: don't know, think
1: skip it, guess, pronounce it, recognition, ask, sound it out, guess from context, look it up in dictionary
2: combination of two or more of the above

Questions D-V-2: Is there a difference between knowing what a word "says" and knowing what a word "means"? For instance, suppose you were reading and you came to a word that you didn't know and you had to figure it out. Once you figured out what it said, would you know what the word meant?

Scoring:

0: no difference, neither
1: difference but no or irrelevant reason
2: difference plus reason, may not know meaning, never seen the word before, hard word

Question D-V-3: Is it better to sound out a word that you don't know or to ask someone what it says? Why?

Scoring:

0: ask, no reason, easier, faster, so that you will be sure, might make mistake sounding out
1: sound it out, no reason, so you won't bother anyone, easier, better retention, learn how to sound others

2: (a) combination of strategies, (try one and then try another if that doesn't work, no reason given) or (b) just ask for meaning (c) combination of strategies, reason given.

Question D-V-4: What would you do if you were reading and you came to a whole sentence that you couldn't understand? [Pause for response.] Is there anything else that you could do?

Scoring:

0: don't know, anything that would help, write, stop

1: ask somebody, skip it, wouldn't happen in English, spell, read aloud, go slowly, try to figure out the words, sound them out, repetition, keep going over it one word at a time, figure out what it means using dictionary, context

2: combination of two of the above

APPENDIX F

Analyses of Other Measures from the Reading Study

Method

In addition to the basic measures for the reading study described in Chapter 4, other sets of measures were obtained in certain conditions of the study. In the Fun instructional condition, children were asked to rate the passage for how good it was on a scale of 1 to 10. This measure was used to verify further the appropriateness of the passages to the ages of the children. In the Title condition, the children were asked to produce an appropriate title for the passage. These titles were judged independently for thematic appropriateness by adult raters, and the resulting scores were used as an indication of the children's ability to identify theme. In the Skim condition, the children were provided with a special question before reading the passage and asked for the answer after reading. This special question also was asked in the other instructional conditions, but in these cases the question was not told to the child until after the child had read the passage, and it was called an "extra question." In addition, reading time was recorded for each passage by use of a stop watch.

Results and Discussion

Fun Ratings

The analysis of the fun ratings was a 2 (grade) $\times$ 3 (reading ability) $\times$ 2 (difficulty level) mixed analysis of variance. The third-grade children gave higher ratings than the sixth-grade children ($F(1, 138) = 12.16, p < .001$). This result might indicate an overenthusiasm or an inability to differentiate stories on the part of younger children. It also could indicate less variety in reading materials for younger children. In addition, the grade X difficulty interaction was significant

Table F-1. Number of Children Responding Correctly to the "Special" Question in Each Instructional Condition

	Grade 3			Grade 6			
Instr.	Poor	Average	Good	Poor	Average	Good	%
Fun(Easy)	4	10	11	6	8	13	36.1
Fun(Hard)	6	7	11	1	3	6	23.6
Title(E)	2	7	11	7	8	14	34.0
Title(H)	5	4	13	2	4	8	25.0
Skim(E)	15	22	20	16	20	23	80.6
Skim(H)	9	14	18	14	10	20	59.0
Study(E)	5	10	12	5	9	15	38.9
Study(H)	4	7	12	3	4	7	25.7

($F(1,138) = 10.60, p < .005$). The third-grade children seemed to like the hard passages better than the easy passages, while the sixth-grade children rated the easy stories higher than the hard stories ($p < .01$). In any case, all ratings tended to be high, indicating that the stories were appropriate for these children. (Mean for grade 3 = 8.17; mean for grade 6 = 7.25.)

Title Ratings

The title ratings were analyzed in a 2 (grade) x 3 (reading ability) x 2 (difficulty level) mixed analysis of variance. The ability to produce a thematically appropriate title increased with reading ability ($F(2,138) = 18.20, p < .001$). Titles suggested by good readers were more thematically appropriate than those suggested by average readers ($p < .01$), which in turn were better than those suggested by poor readers ($p < .01$). This result adds further support to the notion that comprehension increases with reading ability.

Skim Questions

If the child answered the skim question correctly, he or she was given a score of 1 (0 if incorrect). The frequency of each type of response for each grade and reading level under each condition is presented in Table F-1, along with the appropriate chi-square value and probability level. As can be seen, more children tended to get the question correct in the Skim conditions. In addition, the older/better reader also tended to get the question correct regardless of the instructional condition. These results seem to indicate that the children generally were following the instructions and that the older/better readers performed better even on incidental information.

Reading Time

The analysis of time scores was a 2 (grade) x 3 (reading ability) x 4 (instructional condition) x 2 (difficulty level) mixed analysis of variance. The main effect of the instructional condition was significant ($F(3,414) = 32.46$, $p < .001$). The Newman-Keuls analyses showed that the children read faster in the skim condition than in all other conditions ($p < .01$). In addition, the children read somewhat faster in the Fun than in the Study condition. These results seem to indicate that the children were following the instructions. Length of time also decreased with grade and with reading ability ($F(1,138) = 15.57$, $p < .001$ and $F(2,138) = 9.42$, $p < .001$, respectively). Good readers read faster than average readers, and average readers read faster than poor readers. Difficulty level also interacted with grade and with reading ability ($F(2,138) = 4.19$, $p < .05$). At the third-grade level, average and good readers read easy passages at the same rate. However, in every other case, good readers read faster than average readers, and average readers read faster than poor readers ($p < .01$).

APPENDIX G

Instructions to Subjects (and Sample Story)

I want you to read some stories for me. You can read them silently to yourself. Each time you read, I am going to keep track of how long you take. Sometimes it won't matter how long you take and sometimes it will, but I will tell you whether or not it will be important before you begin each story. Before each story, I will tell you about a special question that I'll ask you after the story. Sometimes the special question will be important and sometimes it won't, but I will tell you if it is important before you begin to read each story. After each story, we also will do some questions on some cards like these. Suppose that the story was like this [show example story, read first sentence]. Then afterwards, the questions might be like this [do sample question]. After each question I'll ask you whether or not you are sure or not sure about your answer. It doesn't really matter whether or not you are sure, I just want to know. Sometimes these questions will be important and sometimes they won't, but I'll tell you before you begin whether or not they are important. So, before each story I'll tell you whether or not you should worry about how long you take, or the special question, or the questions on the cards. OK.? [The child was shown where each potentially special item would be recorded on the answer sheet.]

Fun

This time, the special question is important, and the question will be "Do you think that other boys and girls your age would enjoy reading this story? If you had to give it a mark out of 10, 10 being very good and 1 meaning not good at all, what mark would you give it?" It doesn't matter how long you take or how well you do on the card questions. So, can you tell me what is going to be important this time?

Title

This time, the special question is important, and the question will be "What do you think would be a good title for this story?" It doesn't matter how long you take or how well you do on the card questions. So, can you tell me what is going to be important this time?

Skim

This time, two things are important, the time and the special question. I want you to finish with the story as quickly as you can, and afterwards the special question will be [the special information question listed after the multiple-choice questions for each story]. It doesn't matter how well you do on the card questions. So, can you tell me what is going to be important this time?

Study

This time, the card questions are important. I want you to try to get all the card questions right. It doesn't matter how long you take, and there won't be a special question this time. So, can you tell be what is going to be important this time?

[The instructions were repeated until the child could repeat what was to be important. After specific instructions:] If you come to a word that you don't know, you can ask me what it says and I'll tell you. When you are through with the story and ready for the important test or question, say "stop" out loud.

Sample Story

Polly Peters went to live for the summer with her grandmother in a very old house near the sea. In front of the house there was a big round petunia bed. Polly liked to take her grandmother's watering can and water the petunias.

Polly Peters had a puppy named Patrick Peters—Pat for short—who went almost everywhere that Polly went. Polly also had a red chair, a fishing pole, a pretty new red swim suit, a rag doll named Marianne. But there was something else Polly wanted. She wanted it because Grandmother wanted it. It was a shiny tin fish weather vane to put on the red barn behind the house.

There had been one there before the big storm. Now Grandmother would sometimes look at the barn and say sadly, "I wish we had found our weather vane. Too bad it blew so far away in that hard wind. We have never been able to find it."

Then Polly would say to herself, "I should like to give Grandmother a weather vane on her birthday."

So the next time Polly went to town, she stopped to look in at the store where they had weather vanes. She saw a fish one that was just right. The man in the store said it cost five dollars. But Polly had only a dollar and four pennies in her pocketbook. Polly thought and thought, but she could not think of any way to make five dollars.

Every morning Polly brought in the milk bottles and the newspaper. She set the table and helped dry the dishes. She helped make the beds, too.

When her work was done, Grandmother would smile and say, "Run along, Polly. Make the most of your time."

Polly would laugh. Then she would put on her new red swim suit and take Pat and her fishing pole and walk to the dock nearby. First she would fish. Later she would swim.

As she finished, Polly watched the other children swim. Bobby and Billy, six-year-old twins, were often there. Polly tried to keep an eye on them because they had not learned to swim well. They sometimes went out where the water was too deep for them.

The twin's mother, Mrs. Livingston, painted beautiful pictures. She often came to the beach to paint. Polly liked to talk to her and watch her paint. Polly told her about Grandmother Peters and the shiny tin fish weather vane that she wanted to get for her grandmother as a birthday surprise.

After supper one night, Polly told her grandmother how that day a big wave had caught the twins. It had sent them tumbling over and over on the beach.

"Their mother had better keep an eye on those twins," said Grandmother.

Polly laughed. "Oh, we both do that," she said.

Everyone said that Polly was one of the best swimmers at the beach. When Grandmother heard of this, she said, "You should swim well, Polly. You go swimming every day all summer."

One day, Polly saw Bobby in the water beyond the end of the dock. Suddenly he was gone from sight under a big wave.

Sample Question

1. Polly Peters went to live for the summer with her grandmother
   (a) in a very new house in the city
   (b) in a very old house near the sea
   (c) in a very new house near a lake
   (d) in a very old house in the country

APPENDIX H

Stories and Questions

A Grade 3 Easy Story

Benjie and his big brother Nick watched the moving men. One man, whose name was Joe, said, "The people told us to throw away some things. I think you boys might like some of them." Then the other moving man carried out an old bike and a ball. "Here, boys," he said.

"A bike!" Nick said. "Thank you."

"Thank you," said Benjie. "Thank you for the ball." They waved good-by to the moving men.

Benjie said, "I guess I can have the football."

"Sure," said Nick. "But it's a basketball, not a football. Now don't bother me."

Benjie picked up the ball. He saw his friend Billy.

"Where did you get the volleyball?" Billy asked.

"From some moving men," Benjie answered. "And it's a basketball!"

"I know that!" Billy said. "Throw it to me."

Benjie threw the ball. It hit the ground. Plop! "That ball is no good," said Billy. "It doesn't bounce!"

"It will bounce," said Benjie. "You wait and see!"

Benjie ran back to his brother. Nick was working hard. Benjie didn't want to bother him. But he wanted his basketball to bounce. He could not play with it if it didn't bounce.

"What do you want now?" Nick asked.

"This ball doesn't bounce, Nick," said Benjie.

"It seems all right to me," said Nick. "It just needs air. Put some air in it."

"Air?" Benjie asked.

"Yes, air," said Nick. "It goes right in this little hole. Now don't bother me."

Benjie started blowing into the little hole. He huffed and huffed. He puffed and puffed. But the ball didn't get any bigger. Then he huffed and puffed again. Nick looked up. "What are you doing?" he asked.

"Air," Benjie said. He pointed to the little hole. Nick laughed.

"You have to put air in with a pin!"

"A pin?" Benjie asked.

"Yes, a pin," said Nick. "Now don't bother me. I have to take my bike to the gas station. I need some air, too."

Benjie watched him go. Where could he get a pin? "Hello," said Benjie's friend Nancy. "Where did you get the soccer ball?"

"It's a basketball! I got it from some moving men. Only it doesn't bounce. If I find a pin, it will bounce."

"A pin?" Nancy asked.

"Yes, a pin," Benjie answered. "You put the pin in this little hole. Then you fill the ball with air."

"Where do you get the air?" Nancy asked.

"From the gas station, I think," said Benjie. "Only I don't have a pin."

"I have a pin," said Nancy. "Wait here and I'll get it." Nancy came back. She handed Benjie a pin. "You can keep it," she said.

"Thanks," said Benjie. "I'll see you later."

He ran off down the steet to the gas station. "Good luck with your football," Nancy called after him.

"It's a BASKETBALL!" Benjie called back.

Nick was at the gas station. Benjie watched Nick put air in his tires. "Did you get the pin?" Nick asked.

"Sure," said Benjie. "Here." Nick laughed. He took the pin from Benjie. He showed it to his friends. His friends' names were Pete, Bill, and Jack. Pete, Bill, and Jack all laughed at Benjie's pin.

Questions

Important sentences are numbers 1, 3, 5, 7, 9, 11, and 13.

1. Benjie and his big brother
   (a) Ned stood and watched
   (b) Tom watched the worker
   (c) Nick watched the moving men
   (d) Ted watched as the men worked

2. They waved good-bye to
   (a) the work men
   (b) the moving men
   (c) their father's friends
   (d) the strangers

3. But it's a
   (a) football, not a basketball
   (b) volleyball, not a baseball
   (c) baseball, not a volleyball
   (d) basketball, not a football

4. From some moving men,
   (a) Benjie answered
   (b) Bert answered
   (c) Bill said
   (d) Bruce stated

5. It doesn't
   (a) work!
   (b) bounce!
   (c) look very good!
   (d) feel clean!

6. "What do you want
   (a) me to do?" Ned demanded
   (b) for it?" Nathan asked
   (c) now?" Nick asked
   (d) anyhow?" Ben demanded

7. It just needs
   (a) air
   (b) a patch
   (c) another hole
   (d) mending

8. Now don't
   (a) follow me
   (b) get in my way
   (c) go away
   (d) bother me

9. You have to
   (a) put water in the ball!
   (b) put air in with a pin!
   (c) put the patch over the hole!
   (d) make a hole for the air to go in!

10. A pin?
   (a) Lucy demanded
   (b) Linda said
   (c) Nancy asked
   (d) Joan asked

11. "I have a
    (a) pin," said Nancy
    (b) pump," said Jill
    (c) knife," answered Linda
    (d) basketball," answered Linda

12. I'll see you
    (a) in the park
    (b) at school
    (c) at home
    (d) later

13. Nick was
    (a) at the gas station
    (b) getting angry
    (c) at the park
    (d) with his friends

14. His friends' names were
    (a) Bert, Pat, and Joe
    (b) Jim, John, and Tom
    (c) Pete, Bill, and Jack
    (d) Jerry, Dave, and Tim

Special Question: What was the name of one of the moving men?

A Grade 3 Hard Story

In front of Tom's house stood a big old elm tree. The house was old and the tree was old—and they went very well together.

All the other houses on Tom's street were brand new, and the trees that stood in front of the houses were brand new, too. They were saplings. It would take many years before the saplings would be as big as Tom's tree.

Tom liked his tree, and the other boys and girls on the street liked it, too. When they played games like hide-and-seek, the big tree was always "home."

In summer, a bird family always made a nest in the elm tree, and Bobby, the squirrel, made it his home the year round. The big old steps across the front of Tom's house stayed cool all summer because the elm tree kept the sun from shining on them. When the first snow fell in winter, how beautiful the tree looked with soft snow on each branch.

One morning in the fall, Tom came out of his house and saw a big green X on the elm tree.

Tom and his dog looked at the X. "That's funny," said Tom. "It wasn't there before. Who did that, Butch?"

Butch wagged his tail and put out his tongue.

"What's it for, Butch?" Tom asked. Butch just wagged his tail some more.

"Come on, Butch. Let's find out," said Tom. He hopped on his new two-wheeler and down the steet he went with Butch following after.

A painter was painting Mrs. Hill's house. "Paint," thought Tom. "Maybe the painter did it. I'll ask him."

But when he did, the painter laughed. "I'm a house painter, not a tree painter," he said. "Besides, I have only white paint, no green paint."

Tom looked inside the painter's pail. It was true. There was no green paint anywhere.

Tom and Butch went on down the street. Suddenly they smelled something good. What a wonderful smell! They followed it to Mrs. Turner's back door.

"Hello, Tom," called Mrs. Turner. "You're just in time for a peanut cookie. And one for Butch, too."

"Thank you, Mrs. Turner," said Tom.

"Bow-wow," said Butch, who knew how to say thank you, too.

"Mrs. Turner, do you know who put a big green X on my elm tree?" asked Tom.

"A big green X?" Mrs. Turner wondered. "I don't know anything about big green X's. I only know about big brown cookies," she laughed.

"Well, here we go again," Tom said to Butch. "We haven't found out yet."

On the sidewalk they met Linda and her doll. "Linda, do you know who put a big green X on my elm tree?'

"Show me," said Linda. Back to the tree they went. "It's a big green X," said Linda.

"I know that, silly. But who put it there?" asked Tom.

"How can I know that if I'm silly?" said Linda. And off she went with her nose in the air to play dolls with her friend Sue.

"Girls!" said Tom. He put his bike beside the tree and called out to Linda. "I didn't mean to make you angry, Linda," he said.

Questions

Important sentences are numbers 1, 3, 4, 7, 10, 12, and 13.

1. In front of Tom's house stood
   (a) a small oak tree
   (b) a little sapling
   (c) a big old elm tree
   (d) a tall old oak tree

2. They were
   (a) small
   (b) saplings
   (c) babies
   (d) shrubs

3. Tom liked his tree and the other boys and girls on the street
   (a) played under it
   (b) didn't mind the tree
   (c) thought it was O.K.
   (d) liked it, too

4. One morning in the fall, Tom came out of his house and saw
   (a) a big green X on the elm tree
   (b) that the tree was gone
   (c) some men under the tree
   (d) a red circle on the oak tree

5. Butch wagged his tail and
   (a) ran down the street
   (b) put out his tongue
   (c) barked playfully
   (d) waited patiently

6. Butch just
   (a) sat there
   (b) waited until he was through
   (c) wagged his tail some more
   (d) looked back at him

7. Tom looked
   (a) at the painter
   (b) at the painter's brush
   (c) in the painter's truck
   (d) inside the painter's pail

8. What a wonderful
   (a) sight
   (b) smell
   (c) taste
   (d) feeling

9. "Mrs. Turner, do you know who put a big green X on
   (a) my oak tree?" demanded Jack
   (b) my maple tree?" cried John
   (c) my elm tree?" asked Tom
   (d) my pine tree?" said Tommy

10. A big green X?
    (a) Mrs. Turner wondered
    (b) Mrs. Thompson asked
    (c) Mrs. Thomas demanded
    (d) Mrs. Taylor said

11. "Linda, do you know who put
    (a) that mark on my oak tree?"
    (b) that paint on my maple tree?"
    (c) the circle on my pine tree?"
    (d) a big green X on my elm tree?"

12. "How can I know that if
    (a) I'm silly?" said Linda
    (b) I wasn't here?" asked Lucy
    (c) I'm dumb?" demanded Karen
    (d) I'm not as smart as you are?" cried Carol

13. Girls!
    (a) cried John
    (b) laughed Jack
    (c) said Tom
    (d) said Ted

14. He put his bike
    (a) beside the tree
    (b) by the car
    (c) in the garage
    (d) near the door

Special Question: What was the squirrel's name?

A Grade 6 Easy Story

Good King Justin had a beautiful kingdom that stretched from the mountains to the sea. In it were dense forests and rich farms. Along the shore the fishermen sang as they brought in boatloads of fish every day. But the king was not happy. He had one great sorrow—his only son was not content to stay at home.

Let the prince hear of a land that no one had explored and he would be off at once to see it. Let him hear of a fierce beast that no one could capture and he would not rest until he had dragged it home. The prince was called "Harold the Daring."

Not far from King Justin's shore was a rocky island ruled by a wicked magician whose name was Duke Rollo.

Whenever the day was clear, the duke would sit in his watchtower with one eye at his spyglass, peering greedily across the water at King Justin's land.

"Ah," he would mutter, "what a wealthy country that must be. How I wish I could get it for myself."

But he could not. For although he had many soldiers, King Justin had still more. And although he had a great deal of magic power, he could use it only on his own rocky island or in the sea that surrounded it.

"There must be some way of getting rid of King Justin and his son without going to war," Duke Rollo said to himself. He frowned as he considered one plan after another. "I think I should go over there and find out what they are like. That may give me a good idea."

So he packed his bag with everything he might need—and down at the very bottom he hid his book of magic. The next morning he dressed himself as a farmer, and while it was still dark, he took a small boat and rowed over to the mainland. He pulled the boat onto the shore and tied it under a clump of willows. Then he started walking to the city.

All along the way the houses were decorated with flags. "Hm," said Duke Rollo to himself, "I wonder why there are so many flags?" But as no one was up yet, he could not ask.

The sun was just rising when the duke reached the castle gate. The gate was locked and guarded by two huge dogs wearing handsome gold collars with their names, Bruno and Juno, in red rubies. But the dogs were asleep.

Duke Rollo took out his magic key and twisted it. But he had forgotten that he could not use his magic power beyond his own rocky island. He became very angry and he shook the gate with all his might.

The two dogs woke up and leaped against the gate.

"Good old Bruno," said the duke, holding out his hand. But Bruno growled and snapped at him. Juno curled her lips back and showed her sharp teeth. She barked fiercely.

Duke Rollo moved back. "I'll fix you!" he cried, picking up a thick stick.

But just then he heard a soft voice saying, "Are you a stranger?"

Questions

Important sentences are numbers 1, 4, 5, 7, 10, 12, and 14.

1. Good King Justin had a beautiful kingdom that
   (a) lay between two mountains
   (b) stretched from sea to sea
   (c) stretched from the mountains to the sea
   (d) lay between two huge rivers

2. In it were
   (a) rich people and poor people
   (b) dense forests and rich farms
   (c) happy people and rich lands
   (d) rich lands and sad people

3. Along the shore the fishermen sang as they
   (a) put their boats into the water
   (b) mended their nets
   (c) looked after the day's catch
   (d) brought in boatloads of fish every day

4. He had one great sorrow—
   (a) his only son was not content to stay at home
   (b) his only daughter was not very beautiful
   (c) his only daughter was not married
   (d) his only son was always going off to war
5. Not far from King Justin's shore was a rocky island ruled by
   (a) a wise magician whose name was Duke Marlow
   (b) a wicked magician whose name was Duke Rollo
   (c) a wicked prince whose name was Rowland
   (d) a wise wizard whose name was Patron
6. He frowned as he considered
   (a) what might happen
   (b) the news that had just reached him
   (c) one plan after another
   (d) the new plan
7. I think I should go over there and find out what
   (a) they are like
   (b) they are up to
   (c) they want
   (d) they are going to do
8. He pulled the boat onto the shore and
   (a) tied it to a post
   (b) began to walk down the beach
   (c) began to walk toward the fishermen
   (d) tied it under a clump of willows
9. Then he started walking
   (a) to the farm
   (b) to the city
   (c) to the dock
   (d) to his boat
10. The gate was locked and guarded by two huge dogs wearing handsome gold collars with their names,
    (a) Bruto and June, in sparking diamonds
    (b) Zeus and Juno, in red rubies
    (c) Bruno and Juno, in red rubies
    (d) Bruno and Zeus, in sparkling diamonds
11. He turned the key and
    (a) twisted it
    (b) walked in
    (c) pushed on the door
    (d) waited

12. But he had forgotten that he could
    (a) not always win
    (b) use his magical power only at certain times
    (c) use his magical tricks only within his own rich kingdom
    (d) not use his magic power beyond his own rocky island

13. She barked
    (a) fiercely
    (b) quietly
    (c) loudly
    (d) softly

14. But just then he heard a soft voice saying,
    (a) "Are you supposed to be here?"
    (b) "Who are you?"
    (c) "Are you a stranger?"
    (d) "Why did you come here?"

Special Question: What was the Duke's name?

A Grade 6 Hard Story

It was late in the afternoon when the mountian lioness knew that the time was near for her cubs to be born. She had borne cubs every other year since her own third year, and these cubs would be her fourth litter. The hillside cave where she would bear and raise them was the same cave she had used before.

But this time, there was one important difference. In the valley two miles below, the old cabin and barn were no longer deserted buildings. Three people now lived in the cabin, and the old barn behind it was full of livestock. The newcomers had been there since the previous fall.

At first their presence had made the lioness uneasy. In those mountains of Idaho, more than thirty miles from the nearest town, people were the rare animals. And the lioness knew that people were the dangerous animals. They were more dangerous than grizzlies or rattlesnakes—the only other living things she had to be wary of.

For years men with dogs and guns had been hunting down mountain lions until only the wariest of these big cats survived.

So now the expectant lioness had stayed away from the valley, alert for any sign that the newcomers were invading her hillside. But they did not do so, and little by little she had lost her uneasiness about them.

Now safe in the hillside cave, she gave birth to the cubs. The first two were males and came along within five minutes of each other. Each was a little less than a foot long, weighed nearly a pound, and was enclosed in a thin bluish membrane.

For several minutes after they were born, the lioness worked at them with her tongue. She licked the membranes off their bodies and off their heads. As the head of each kitten was freed, he drew his first breath and uttered a faint mew.

Then the third and last of the cubs was born. She was about half the size of her brothers, a little less than half a pound and barely six inches long. She was the runt of the litter, unusually small, even for a female cub.

The eyes of all three cubs were closed and would remain so for nine or ten days, but the two males easily found their mother's warm underside and fed greedily while the mother was still cleaning the littlest cub. Noticeably less strong than her brothers, it took her much longer to find her way to a nipple.

During the next several days, life for the cubs was drinking their mother's milk, sleeping, and being cleaned by her tongue. Differences in strength between the female cub and her two brothers grew greater. They fed more often and sometimes pushed her away when she tried to nurse near one of them.

The mother did not try to help the littlest one. Lion mothers do not notice when one cub is not getting enough to eat. Also, this mother soon began running short of milk for her litter.

The weather was very cold and she could not leave the cubs alone for long. They were still so helpless they would freeze to death without her to keep them warm.

Questions

Important sentences are numbers 1, 3, 4, 8, 9, 13, and 14.

1. It was late in the afternoon when the mountain lioness knew that
   (a) danger was near
   (b) the time had come when something must be done
   (c) the time was near for her cubs to be born
   (d) she must escape

2. She had borne cubs every other year since her own third year, and
   (a) she was not afraid at all
   (b) these cubs would be her fourth litter
   (c) she knew exactly what to expect
   (d) she knew what she must do

3. The hillside cave where she would bear and raise them was
   (a) close to fresh water and food
   (b) near the same cave where she was born
   (c) lined with soft leaves
   (d) the same cave she had used before

4. Three people now lived in the cabin, and the old barn behind it was
   (a) full of livestock
   (b) being repaired
   (c) being torn down
   (d) still empty

5. At first their presence had made
   (a) no difference at all
   (b) the lioness uneasy
   (c) little difference
   (d) the lioness happy

6. In those mountains of Idaho, more than thirty miles from the nearest town, people were
   (a) always coming to hunt
   (b) building summer homes
   (c) the rare animals
   (d) very rare

7. They were more dangerous than grizzlies or rattlesnakes—
   (a) the only other living things she had to be wary of
   (b) the only things that she did not try to capture
   (c) the only other living things that were a threat to her
   (d) the only living things that were a threat to these humans

8. Now safe in the hillside cave, she
   (a) began to relax
   (b) began to get worried
   (c) looked after her cubs
   (d) gave birth to the cubs

9. The first two were males and
   (a) were about the same size as each other
   (b) came along within five minutes of each other
   (c) came along about one hour apart
   (d) were very different in size

10. Each was a little less than a foot long, weighed nearly a pound, and
    (a) was blind at birth
    (b) was bluish at birth
    (c) was enclosed in a thin bluish membrane
    (d) was covered with a sticky white substance

11. She licked the membranes off their bodies and
    (a) off their heads
    (b) off their eyes
    (c) pushed them away
    (d) then left them alone

12. As the head of each kitten was freed, he
    (a) started to cry and looked around
    (b) looked around and tried out his legs

(c) stretched and started to mew loudly
(d) drew his first breath and uttered a faint mew

13. She was the runt of the litter,
    (a) unusually small, even for a female cub
    (b) unusually large, even for a female cub
    (c) unusually awkward, even for a newborn
    (d) unusually small, even for the first born

14. Differences in strength between the female cub and her two brothers
    (a) grew less
    (b) became less noticeable
    (c) grew greater
    (d) began to bother the mother

Special Question: Where were the mountains?

References

Bridwell, N. *The cat and the bird in the hat*. Toronto: Scholastic Book Services, 1964.

Bridwell, N. Clifford, *the small red puppy*. Toronto: Scholastic Book Services, 1972.

Cebulash, M. *The ball that wouldn't bounce* (pp. 5-20). Toronto: Scholastic Book Services, 1972.

Cleaver, N. *How the Chipmunk got its stripes* (pp. 1-8). Toronto: Scholastic-Tab Publications, Ltd., 1973.

Clymer, E. *The big pile of dirt* (pp. 5-12). Toronto: Scholastic Book Services, 1968.

Cook, M. *Waggles and the dog catcher* (pp. 3-17). Toronto: Scholastic Book Services, 1963.

De Angeli, M. *The door in the wall* (pp. 86-87). Toronto: Scholastic Book Services, 1949.

Froman, R. *The wild orphan* (pp. 5-8, 10-12). Toronto: Scholastic Book Services, 1972.

Hodges, M. *What's for lunch, Charley* (pp. 3-10). Toronto: Scholastic Book Services, 1961.

Myers, B. *Not THIS bear!* (pp. 1-31). Toronto: Scholastic Book Services, 1967.

Nevin, E. *The estraordinary adventures of Chee Chee McNerney* (pp. 103-105). Toronto: Scholastic Book Services, 1971.

Nielsen, V. *The house on the volcano* (pp. 5-7). Toronto: Scholastic Book Services, 1966.

Piper, E. *The Magician's trap* (pp. 1-3). Toronto: Scholastic-Tab Publications, 1976.

Serraillier, I. *The Gorgon's head: The story of Perseus* (pp. 1-3). Toronto: Scholastic Book Services, 1961.

Woyke, C. *Goodby, tree!* (pp. 3-8). New York, NY: MacMillan Co., 1967.

APPENDIX I

Passage-Controlled Comparisons

Initial scanning of the experimental comprehension scores indicated that the third-grade good readers were performing at or above the reading level of sixth-grade poor readers. However, since the children were reading grade appropriate stories, it was impossible to tell if this relationship was "real" or a result of the difficulty of the passages. Fortunately, the categorization of level of difficulty was such that the difference in difficulty level between the two sets of stories at each grade was the same as the difference between the hard stories at the third-grade level and the easy stories at the sixth-grade level. Given this fortuitous circumstance, it was decided that these two groups would do a third session with a third level of "difficulty." In this session, the third-grade good readers read the sixth-grade easy passages and the sixth-grade poor readers read the third-grade hard passages. The procedure was the same as that used for the first two sessions, except that in this session, each child read one passage for each goal instruction, and all four goal instructions were covered in one session. This third reading session was given after all the other sessions had been completed.

Results and Discussion

The following analyses were done using the results from the same stories.

Comprehension

The analysis was a 2 (group) x 4 (instructional condition) x 2 (difficulty level) x 2 (importance of sentence) mixed analysis of variance. Thematically important sentences were remembered better than less important sentences ($F(1,46) = 85.61$, $p < .001$), and comprehension decreased as difficulty level increased ($F(1,46) = 26.28$, $p < .001$). In addition, the main effect of the instructional condition was significant ($F(3,138) = 8.02$, $p < .001$). Comprehension was lower in the Skim condition than in all other instructional conditions ($p < .01$). The interaction of importance of sentence and difficulty level also was significant ($F(1,46) = 17.71$, $p < .001$), and important sentences were retained in the easy passages more frequently than in the hard passages ($p < .01$). In addition, less important sentences were retained slightly better in easy than in hard passages. There was no difference between the two grades ($F(1,46) = 0.01$, $p < .10$). These results would seem to indicate that third-grade good and sixth-grade poor readers were comprehending at the same level.

Prediction Accuracy

The analysis was a 2 (group) x 4 (instructional condition) x 2 (difficulty level) x 2 (importance of sentences) mixed analysis of variance. Thematically important sentences were easier to predict than less important sentences ($F(1,46) = 33.96$, $p < .001$). In addition, importance of sentence interaacted with difficulty level ($F(1,46) = 7.50$, $p < .01$). Important sentences were predicted more accurately in easy than in hard passages ($p < .01$). The difference between important and less important sentences was greater in the easy passages ($p < .01$) than in the hard passages. Success at prediction also was affected by instructional condition ($F(3,138) = 3.81$, $p < .05$). Accuracy of predictions was lower in the skim than in either the title or the study conditions. Once again there was no grade difference ($F(1,46) = 0.59$, $p < .10$). It would appear, then, that in terms of ability to predict comprehension accuracy, third-grade good and sixth-grade poor readers were at about the same level.

APPENDIX J

Source Tables for the Reading Study

List of Tables

J-27: Grade 3 Good—Grade 6 Poor Fun Ratings: ANOVA
J-28: Grade 3 Good—Grade 6 Poor Rate of Reading: ANOVA
Note: In all cases, IQ refers to a nonverbal measure of intelligence.

Table J-1. Comprehension: ANOVA

Source	df	MS	F	*p*
Sentence Type (S)	1	286.17	179.31	.001
Difficulty (D)	1	18.42	10.19	.005
Instruction (I)	3	40.35	13.60	.001
Reading Ability (R)	2	585.63	47.86	.001
Grade (G)	1	321.01	26.23	.001
SG	1	12.54	7.86	.01
DRG	2	10.16	5.62	.005
DIRG	6	5.04	2.71	.05
Error Terms				
S/Groups	138	12.24		
S/S	138	1.60		
S/D	138	1.81		
S/S&D	138	1.26		
S/I	414	2.97		
S/S&I	414	1.41		
S/D&I	414	1.86		
S/S&D&I	414	1.37		

Note: Only significant source items are listed in this table.

Table J-2. Comprehension: COVA

Source	df	MS	F	*p*
Grade (G)	1	330.57	27.74	.001
Reading Ability (R)	2	375.03	31.47	.001
IQ	1	56.02	4.70	.05
Error	137	11.92		
Instruction (I)	3	40.35	13.60	.001
Error	414	2.97		
Difficulty (D)	1	18.42	10.19	.005
DGR	2	10.16	5.62	.005
Error	138	1.81		
IDGR	6	5.04	2.71	.05
Error	414	1.86		
Sentence Type (S)	1	286.17	179.31	.001
SG	1	12.54	7.86	.01
Error	414	1.60		

Note: Only significant source items are listed in this table.

Table J-3. Grade 3 Poor Readers' Comprehension: ANOVA

Source	df	MS	F	*p*
Sentence Type (S)	1	41.34	23.98	.001
Error Term				
S/S	23	1.72		

Note: Only significant source items are listed in this table.

Table J-4. Grade 3 Poor Readers' Comprehension: COVA

Source	df	MS	F	*p*
Sentence Type (S)	1	41.34	23.98	.001
Error	23	1.72		

Note: Only significant source items are listed in this table.

Table J-5. Grade 3 Average Readers' Comprehension: ANOVA

Source	df	MS	F	*p*
Sentence Type (S)	1	66.67	30.09	.001
Error Term				
S/S	23	2.22		

Note: Only significant source items are listed in this table.

Table J-6. Grade 3 Average Readers' Comprehension: COVA

Source	df	MS	F	*p*
Sentence Type (S)	1	66.67	30.09	.001
Error	23	2.22		

Note: Only significant source items are listed in this table.

Table J-7. Grade 3 Good Readers' Comprehension: ANOVA

Source	df	MS	F	*p*
Sentence Type (S)	1	109.44	65.81	.001
Instruction (I)	3	18.71	4.77	.01
SDI	3	3.24	3.75	.05
Error Terms				
S/S	23	1.66		
S/I	69	3.93		
S/S&D&I	69	0.86		

Note: Only significant source items are listed in this table.

Table J-8. Grade 3 Good Readers' Comprehension: COVA

Source	df	MS	F	*p*
Instruction (I)	3	18.71	4.77	.005
Error	69	3.93		
Sentence Type (S)	1	109.44	65.81	.001
Error	23	1.66		
IDS	3	3.24	3.75	.05
Error	69	0.86		

Note: Only significant source items are listed in this table.

Table J-9. Grade 6 Poor Readers' Comprehension: ANOVA

Source	df	MS	F	*p*
Sentence Type (S)	1	47.46	35.10	.001
Difficulty (D)	1	16.25	6.67	.05
Instruction (I)	3	16.94	6.26	.001
Error Terms				
S/S	23	1.35		
S/D	23	2.44		
S/I	69	2.70		

Note: Only significant source items are listed in this table.

Table J-10. Grade 6 Poor Readers' Comprehension: COVA

Source	df	MS	F	*p*
Instruction (I)	3	16.94	6.26	.001
Error	69	2.70		
Difficulty (D)	1	16.25	6.67	.05
Error	23	2.44		
Sentence Type (S)	1	47.46	35.10	.001
Error	23	1.35		

Note: Only significant source items are listed in this table.

Table J-11. Grade 6 Average Readers' Comprehension: ANOVA

Source	df	MS	F	*p*
Sentence Type (S)	1	15.84	10.44	.005
Difficulty (D)	1	15.04	8.82	.01
Error Terms				
S/S	23	1.52		
S/D	23	1.70		

Note: Only significant source items are listed in this table.

Table J-12. Grade 6 Average Readers' Comprehension: COVA

Source	df	MS	F	*p*
Difficulty (D)	1	15.04	8.82	.01
Error	23	1.70		
Sentence Type (S)	1	15.84	10.44	.005
Error	23	1.52		

Note: Only significant source items are listed in this table.

Table J-13. Grade 6 Good Readers' Comprehension: ANOVA

Source	df	MS	F	*p*
Sentence Type (S)	1	30.38	27.53	.001
Instruction (I)	3	13.75	9.78	.001
Error Terms				
S/S	23	1.10		
S/I	69	1.41		

Note: Only significant source items are listed in this table.

Table J-14. Grade 6 Good Readers' Comprehension: COVA

Source	df	MS	F	*p*
Instruction (I)	3	13.75	9.78	.001
Error	69	1.41		
Sentence Type (S)	1	30.37	27.53	.001
Error	23	1.10		

Note: Only significant source items are listed in this table.

Table J-15. Efficiency Scores: ANOVA

Source	df	MS	F	*p*
Sentence Type (S)	1	54.43	169.04	.001
Difficulty (D)	1	3.80	11.90	.001
Instruction (I)	3	4.24	7.96	.001
Reading Ability (R)	2	153.71	80.93	.001
Grade (G)	1	106.98	56.32	.001
SG	1	1.55	4.83	.05
DRG	2	1.25	3.93	.05
DIRG	6	1.12	3.05	.01
Error Terms				
S/GROUPS	138	1.90		
S/S	138	0.32		
S/D	138	0.32		
S/S&D	138	0.25		
S/I	414	0.53		
S/S&I	414	0.27		
S/D&I	414	0.37		
S/S&D&I	414	0.27		

Note: Only significant source items are listed in this table.

Table J-16. Efficiency Scores: COVA

Source	df	MS	F	*p*
Grade (G)	1	1577.27	51.35	.001
Reading Ability (R)	2	459.77	14.97	.001
Error	137	30.71		
Instruction (I)	3	51.39	6.68	.001
Error	414	7.69		
DGR	2	30.13	5.97	.005
Error	138	5.05		
Sentence Type (S)	1	311.37	55.41	.001
Error	138	5.62		
IS	3	12.95	3.40	.05
Error	414	3.80		

Note: Only significant source items are listed in this table.

Table J-17. Prediction Accuracy: ANOVA

Source	df	MS	F	*p*
Sentence Type (S)	1	311.38	55.41	.001
Instruction (I)	3	51.39	6.68	.001
Reading Ability (R)	2	753.44	24.09	.001
Grade (G)	1	1548.75	49.52	.001
SI	3	12.95	3.41	.05
DRG	2	30.13	5.97	.01
Error Terms				
S/GROUPS	138	31.27		
S/S	138	5.62		
S/D	138	5.05		
S/I	414	7.69		
S/S&I	414	3.80		

Note: Only significant source items are listed in this table.

Table J-18. Prediction Accuracy: COVA

Source	df	MS	F	*p*
Grade (G)	1	1577.27	51.35	.001
Reading Ability (R)	2	459.77	14.97	.001
Error	137	30.71		
Instructions (I)	3	51.39	6.68	.001
Error	414	7.69		
DGR	2	30.13	5.97	.005
Error	138	5.05		
Sentence Type (S)	1	311.37	55.41	.001
Error	138	5.62		
IS	3	12.95	3.40	.05
Error	414	3.80		

Note: Only significant source items are listed in this table.

Table J-19. Title Ratings: ANOVA

Source	df	MS	F	*p*
Reading Ability (R)	2	4560.79	18.20	.001
Error	138	250.62		

Note: Only significant source items are listed in this table.

Table J-20. Fun Ratings: ANOVA

Source	df	MS	F	*p*
Grade (G)	1	61.42	12.16	.001
DG	1	31.34	10.60	.005
Error Terms				
S/GROUPS	138	5.05		
S/D	138	2.96		
Total	287	4.29		

Note: Only significant source items are listed in this table.

Table J-21. Rate of Reading: ANOVA

Source	df	MS	F	*p*
Grade (G)	1	4.03	15.56	.001
Reading Ability (R)	2	2.44	9.42	.001
Error	138	0.26		
Instructions (I)	3	0.90	32.46	.001
Error	414	0.03		
DGR	2	0.03	4.19	.05
Error	138	0.01		

Note: Only significant source items are listed in this table.

Table J-22. Grade 3 Good–Grade 6 Poor Comprehension: ANOVA

Source	df	MS	F	*p*
Sentence Type (S)	1	130.85	85.61	.001
Difficulty (D)	1	50.53	26.28	.001
Instruction (I)	3	21.80	8.02	.001
SD	1	16.04	17.71	.001
Error Terms				
S/S	46	1.53		
S/D	46	1.92		
S/S&D	46	0.91		
S/I	138	2.72		

Note: Only significant source items are listed in this table.

Table J-23. Grade 3 Good–Grade 6 Poor Comprehension: COVA

Source	df	MS	F	*p*
Instruction (I)	3	21.80	8.01	.001
Error	138	2.72		
Difficulty (D)	1	50.53	26.28	.001
Error	46	1.92		
Sentence Type (S)	1	130.84	85.60	.001
Error	46	1.53		
DS	1	16.04	17.71	.001
Error	46	0.91		

Note: Only significant source items are listed in this table.

Table J-24. Grade 3 Good–Grade 6 Poor Prediction Accuracy: ANOVA

Source	df	MS	F	p
Sentence Type (S)	1	142.66	33.96	.001
Instruction (I)	3	22.14	3.38	.05
SD	1	35.45	7.50	.01
Error Terms				
S/S	46	4.20		
S/S&D	46	4.72		
S/I	138	6.55		

Note: Only significant source items are listed in this table.

Table J-25. Grade 3 Good–Grade 6 Poor Prediction Accuracy: COVA

Source	df	MS	F	p
IQ	1	140.82	5.22	.05
Error	45	26.99		
Instruction (I)	3	22.14	3.38	.05
Error	138	6.55		
Sentence Type (S)	1	142.66	33.96	.001
Error	46	4.20		
DS	1	35.45	7.50	.01
Error	46	4.72		

Note: Only significant source items are listed in this table.

Table J-26. Grade 3 Good–Grade 6 Poor Title Ratings: ANOVA

Source	df	MS	F	p
Difficulty (D)	1	2470.51	18.85	.001
Error (S/D)	46	131.04		

Note: Only significant source items are listed in this table.

Table J-27. Grade 3 Good–Grade 6 Poor Fun Ratings: ANOVA

Source	df	MS	F	p
Difficulty (D)	1	2.67	0.70	n.s.
Grade (G)	1	1.50	0.24	n.s.
DG	1	12.04	3.16	n.s.
Error Terms				
S/GROUPS	46	6.24		
S/D	46	3.81		
Total	95	5.04		

Table J-28. Grade 3 Good–Grade 6 Poor Rate of Reading: ANOVA

Source	df	MS	F	*p*
Instruction (I)	3	0.21	17.46	.001
Error	138	0.01		
DG	1	0.10	6.80	.01
Error	46	0.01		

Note: Only significant source items are listed in this table.

APPENDIX K

Comprehension and Strategies: Summary of Statistical Analyses

Table K-1. Comprehension and Strategies Performance: Significant Effects

Grade	$F(1,138) = 26.23, p < .001$
Reading Ability	$F(2,138) = 47.86, p < .001$
Instructional Condition	$F(4,414) = 13.60, p < .001$
Difficulty Level	$F(1,138) = 10.19, p < .005$
Importance of Sentence	$F(1,138) = 179.31, p < .001$
Grade × Reading Level × Instruction Condition × Difficulty Level	$F(6,414) = 2.71, p < .05$
Group Analyses of Four-Way Interaction	
Sixth-grade good readers	
Importance of Sentence	$F(1,23) = 27.53, p < .001$
Instructional Condition	$F(3,69) = 9.78, p < .001$
Sixth-grade average readers	
Importance of Sentence	$F(1,23) = 10.44, p < .001$
Difficulty Level	$F(1,23) = 8.82, p < .01$
Instructional Condition	$F(3,69) = 2.38, p < .05$
Sixth-grade poor readers	
Importance of Sentence	$F(1,23) = 35.10, p < .001$
Difficulty Level	$F(1,23) = 6.67, p < .05$
Instructional Condition	$F(3,69) = 6.26, p < .001$
Third-grade good readers	
Importance of Sentence	$F(1,23) = 65.81, p < .001$
Instructional Condition	$F(3,69) = 4.77, p < .01$
Instructional Condition × Importance of Sentence × Difficulty Level	$F(3,69) = 3.70, p < .05$
Third-grade average readers	
Importance of Sentence	$F(1,29) = 30.09, p < .001$
Third-grade poor readers	
Importance of Sentence	$F(1,23) = 23.98, p < .001$
Other Interactions	
Grade × Importance of Sentence	$F(1,138) = 7.86, p < .01$
Grade × Reading Ability × Difficulty Level	$F(2,138) = 5.62, p < .005$
Analysis of Covariance	
Grade	$F(1,137) = 27.24, p < .001$
Reading Ability	$F(2,137) = 31.47, p < .001$
Grade × Reading Ability × Difficulty Level × Instructional Condition	$F(6,414) = 2.71, p < .05$

Table K-2. Comprehension Performance: Computed Scores

Grade	$F(1,138) = 26.17, p < .001$
Reading Ability	$F(2,138) = 48.22, p < .001$

Table K-3. Prediction-Accuracy Scores: Significant Effects

Grade	$F(1,138) = 49.52, p < .001$
Reading Ability	$F(2,138) = 24.09, p < .001$
Importance of Sentence	$F(1,138) = 55.41, p < .001$
Instructional Condition	$F(1,138) = 6.68, p < .001$
Instructional Condition × Importance of Sentence	$F(3,414) = 6.68, p < .05$
Difficulty × Grade × Reading Ability	$F(2,138) = 5.97, p < .01$

Table K-4. Comprehension Verbalization: Chi-Square Results

Item	Grade	Reading Ability	Other
C-V-1	$\chi^2(1) = 5.53, p < .05$	n.s.	n.s.
C-V-2	$\chi^2(1) = 24.03, p < .0001$	n.s.	n.s.
C-V-3	$\chi^2(1) = 16.03, p < .0001$	n.s.	n.s.

Table K-5. Comprehension Verbalization: Correlations with Reading Ability

	Grade 3		Grade 6	
Measure	*r*	partial *r*	*r*	partial *r*
C-V-1	.0629	.0877	.2554*	.1698
C-V-2	.2514*	.1553	.2576*	.1547
C-V-3	.1344	.0558	−.0404	−.1045

$*p < .05$

Table K-6. Comprehension Verbalization: Computed Scores

Grade	$F(1,138) = 74.92, p < .001$
Reading Ability	$F(2,138) = 28.55, p < .001$

Table K-7. Comprehension: Meta Categorizations

Overall	$\chi^2(15) = 67.89, p < .0001$
Grade	$\chi^2(1) = 5.50, p < .05$
Reading Ability	$\chi^2(2) = 31.35, p < .0001$
Poor vs. average readers	$\chi^2(1) = 12.20, p < .001$
Third-grade poor vs. average readers	$\chi^2(1) = 4.25, p < .05$
Sixth-grade poor vs. average readers	$\chi^2(1) = 6.94, p < .01$
Third-grade good vs. Sixth-grade good readers	$\chi^2(1) = 5.58, p < .05$

Table K-8. Strategies Verbalization: Chi-Square Results

Item	Grade	Reading	Other
S-V-1	n.s.	n.s.	n.s.
S-V-2	$\chi^2(1) = 16.94$, $p < .0001$	$\chi^2(2) = 19.51$ $p < .005$	poor < average: $\chi^2(1) = 4.67$, $p < .05$ gr.6 poor < gr.6 average $\chi^2(1) = 4.25$, $p < .05$
S-V-3	$\chi^2(1) = 26.28$, $p < .0001$	n.s.	n.s.
S-V-4	$\chi^2(1) = 12.55$, $p < .001$	$\chi^2(2) = 6.87$, $p < .05$	average < poor: $\chi^2(1) = 3.90$, $p < .05$
S-V-5	$\chi^2(1) = 21.16$, $p < .0001$	$\chi^2(2) = 12.34$, $p < .01$	average < good: $\chi^2(1) = 3.90$, $p < .05$
S-V-6	$\chi^2(1) = 31.97$, $p < .0001$	n.s.	n.s.

Table K-9. Strategies Verbalization: Correlations with Reading Ability

	Grade 3		Grade 6	
Measure	*r*	partial *r*	*r*	partial *r*
S-V-1	.1206	.0379	.0565	−.0448
S-V-2	.1262	.0491	.3444**	.2029*
S-V-3	.1871	.1558	.2535*	.1538
S-V-4	.1462	−.0439	.1573	.1000
S-V-5	.0871	.0436	.4356***	.3317**
S-V-6	.3197**	.2183*	−.0013	-.0148

*$p < .05$ **$p < .01$ ***$p < .001$

Table K-10. Strategies: Computed Scores

Verbalization	
Grade	$F(1,138) = 94.58, p < .001$
Reading Ability	$F(2,138) = 12.30, p < .001$
Performance	
Grade	$F(1,138) = 56.05, p < .001$
Reading Ability	$F(2,138) = 80.92, p < .001$

Table K-11. Strategies: Meta Categorizations

Overall	$\chi^2(15) = 102.56, p < .0001$
Grade	$\chi^2(1) = 26.13, p < .0001$
Reading Ability	$\chi^2(2) = 25.61, p < .0001$
Poor vs. average readers	$\chi^2(1) = 10.10, p < .01$
Sixth-grade poor vs. average readers	$\chi^2(1) = 6.80, p < .01$

APPENDIX L

Comprehension: Verbalization Items

Question C-V-1: Suppose I gave you a story to study and told you that there was going to be a test on the story later. You could write the test whenever you were ready. How would you know when you were ready?

Scoring:

0: write it whenever I was done reading, or doesn't matter

1: use of immature strategy such as read it, write it all down, study it (no explanation), when I know the story; use of one strategy alone such as re-read, write down parts, rehearse

2: use of advanced strategy such as self-testing, other-testing, if I could say by heart

Question C-V-2: Suppose one day at school you wrote a test and when you went home, your mom asked you how the test was—was it hard or easy, did you do well—that sort of thing. How would you know how to answer her questions?

Scoring:

0: can't tell before I get test back

1: remember if hard or easy, studied a lot before test

2: indication of reasons for hard/easy such as length of time to finish, number that you are sure of, didn't know words

Question C-V-3: Suppose I brought in a new game that you had never seen before and asked you to teach your friend how to play the game. Could you do it? What would you need to know about the game before you could teach somebody how to play it? [Note: Most children knew they would not be able to teach unless they had read the instructions.] How could you tell when you knew enough about the game to teach someone else how to play?

Scoring:

0: don't know, or would just know, ask someone else
1: read directions then knew it
2: play it once or twice, self-test, read directions then try it once

APPENDIX M

Strategies: Verbalization Items

Question S-V-1: What do you do when you read and you know that there will be a test on the story later?

Scoring:
- 0: don't know, cheat, read fast and remember, irrelevant
- 1: read it over, practice, rehearsal, study (no explanation), slow and careful, try to remember, read out loud
- 2: self-testing, other-testing, notes, concentrate, parts

Question S-V-2: Can you do anything as you read that would make what you are reading easier to remember? [Pause] Anything else?

Scoring:
- 0: nothing, keep in head, cheat
- 1: read it over once, figure out all the words, aloud, slow, careful, write down story, study (no explanation), good or easy story easier to remember, practice, rehearsal, go over and over it, concentrate, break into parts, make images of what is happening
- 2: take notes, remember the main points, other-test, self-test, concentrate for details

Question S-V-3: Suppose I gave you a story to read and asked you to read it so that afterwards you would be able to tell me the name of the place where the people in the story lived. You wouldn't need to remember anything else, just the name of the place. What would you do to find the name of the place? [Pause] Is that the best way to find the name of the place? Is there anything else that you could do?

Scoring:

0: couldn't do it
1: read it all, search, just look, read some parts (any parts), read it til you find name, look for right part, or read whole to find out more
2: skim it till you find name, just look through, don't read it, go quickly, skim all just in case, logical search such as index or look for capital letters

Question S-V-4: Suppose I gave you a story to read and asked you to read it so that you could tell it to your friend later. What would you do so that you could remember the story to tell it to your friend?

Scoring:

0: nothing, just remember it, read once
1: write it down, use book, read again, read slower, memorize, study, relax, understand it, write down parts, make notes, concentrate, rehearsal
2: write down important parts, remember important parts, use different words to make shorter, self-test, other-test

Questions S-V-5: How much of the story do you think you would be able to remember?

Scoring:

0: all or none, most three-quarters, no qualification
1: one-quarter or one-half, specific parts, depends on other variables
2: important parts

Question S-V-6: Suppose I gave you a story without a title and asked you to make up the best title that you could for the story. How would you think of a title?

Scoring:

0: couldn't do it, just read, ask someone
1: think of one, read and think of one
2: thematize

APPENDIX N

Developmental Items: Statistical Analyses

Table N-1. Language Performance Items: Analyses

Overall multivariate analysis
Grade, $F(15,124) = 15.61$, $p < .0001$
Reading Ability, $F(10,246) = 8.40$, $p < .0001$
Grade × Reading Ability, $F(30,246) = 1.55$, $p < .05$

Univariate F's

Item	Grade $df = (1,138)$	Reading $df = (2,138)$	Grade × Reading $df = (2,138)$
L-P-1	11.65***	105.78****	0.46
L-P-2	deleted	—	—
L-P-3	deleted	—	—
L-P-4	15.19***	4.32	0.28
L-P-5	10.83**	1.71	1.68
L-P-6	2.57	1.68	1.68
L-P-7	4.21*	6.84**	0.37
L-P-8	4.52	2.98	0.10
L-P-9	56.69****	5.96**	1.21
L-P-10	7.19**	4.90**	0.58
L-P-11	72.05****	18.83****	2.02
L-P-12	82.41****	14.11****	2.68
L-P-13	62.21****	23.06****	1.97
L-P-14	12.28***	0.90	1.13
L-P-15	1.61	0.62	0.65
L-P-16	6.86**	1.60	2.97

*$p < .05$ **$p < .01$ ***$p < .001$ ****$p < .0001$

Table N-2. Language Performance Items: Correlations with Reading Ability

Measure	Grade 3		Grade 6	
	r	partial r	r	partial r
L-P-1	.8300***	.7982***	.8085***	.7430***
L-P-2	deleted	—	—	—
L-P-3	deleted	—	—	—
L-P-4	.2472*	.1314	.2081*	.1798
L-P-5	.2001*	.1677	−.0626	−.1763
L-P-6	.2189*	.2093*	deleted	—
L-P-7	.3211**	.2524*	.2183*	.1824
L-P-8	.2201*	.2375*	.1929*	.1509
L-P-9	−.1519	−.0312	−.3716***	−.3403**
L-P-10	.2609**	.2169*	.2436*	.2547*
L-P-11	.4020***	.3093***	.1929*	.1509
L-P-12	.4055***	.3472**	.4385***	.3104**
L-P-13	.4842***	.3984***	.4378***	.3649***
L-P-14	−.2499*	−.2347*	.0132	.0272
L-P-15	−.2230*	−.2683*	−.0445	−.0527
L-P-16	−.3847***	−.3384**	.1382	.1673

*$p < .05$ **$p < .01$ ***$p < .001$

Table N-3. Memory Performance Items: Analyses

Overall multivariate analysis
Grade, $F(7,132) = 28.83$, $p < .0001$
Reading Ability, $F(14,262) = 4.24$, $p < .0001$
Grade × Reading Ability, n.s.

Univariate F's

Item	Grade $df = (1,138)$	Reading $df = (2,138)$	Grade × Reading $df = (2,138)$
M-P-1	134.50****	9.62****	3.19*
M-P-2	54.08****	8.73***	0.18
M-P-3	84.67****	7.72***	0.41
M-P-4	66.98****	8.06***	2.66
M-P-5	deleted	—	—
M-P-6	10.07**	1.77	1.14
M-P-7	29.85****	3.62*	0.80
M-P-8	deleted	—	—

*$p < .05$ **$p < .01$ ***$p < .001$ ****$p < .0001$

Table N-4. Memory Performance Items: Correlations with Reading Ability

Measure	Grade 3		Grade 6	
	r	partial *r*	*r*	partial *r*
M-P-1	.3717***	.3325**	.3545***	.1943*
M-P-2	.4127***	.3040**	.3391**	.1441
M-P-3	.2765**	.2220*	.3619***	.1720
M-P-4	.4206***	.3085**	.2913**	.1432
M-P-5	deleted	—	—	—
M-P-6	.0296	−.0093	.2601*	.1400
M-P-7	.1160	.1307	.3144	.1322
M-P-8	deleted	—	—	—

*$p < .05$ **$p < .01$ ***$p < .001$

Table N-5. Attention Performance Items: Analyses

Overall multivariate analysis
Grade, $F(13, 126) = 30.27$, $p < .0001$
Reading Ability, $F(26, 250) = 3.32$, $p < .0001$
Grade × Reading Ability, $F(26, 250) = 2.23$, $p < .001$

Univariate F's

Item	Grade $df = (1, 138)$	Reading $df = (2, 138)$	Grade × Reading $df = (2, 138)$
A-P-1	6.22**	1.67	0.10
A-P-2	5.23*	2.41	0.67
A-P-3	41.84****	7.29***	2.89
A-P-4	31.01****	6.08**	5.28**
A-P-5	53.83****	3.09*	2.49
A-P-6	38.49****	11.10****	4.36*
A-P-7	6.32**	3.00	2.32
A-P-8	131.34****	13.04****	0.16
A-P-9	20.82****	1.47	1.06
A-P-10	40.61****	3.43*	3.48*
A-P-11	87.39****	0.85	1.19
A-P-12	34.85****	2.11	1.33

*$p < .05$ **$p < .01$ ***$p < .001$ ****$p < .0001$

Table N-6. Attention Performance Items: Correlations with Reading Ability

	Grade 3		Grade 6	
Measure	r	partial r	r	partial r
A-P-1	.2003	.1247	.2296*	.1322
A-P-2	.2061*	.1028	.1487	.1747
A-P-3	.4680***	.4253***	.2277*	.1437
A-P-4	.4009***	.3619***	.1341	.0433
A-P-5	.2035*	.0928	.3009**	.0755
A-P-6	.4787***	.4208***	.3034**	.2763**
A-P-7	−.2016*	−.1235	.1305	−.0091
A-P-8	.4193***	.3262**	.3725***	.2796**
A-P-9	.0687	.1041	−.1116	−.2274
A-P-10	.2036*	.1297	.0315	.0726
A-P-11	.0634	.0376	.2388*	.1186
A-P-12	.1494	.1154	.0459	.1070

*$p < .05$ **$p < .01$ ***$p < .001$

Table N-7. Language Verbalization Items: Chi-Square Results

Item	Grade	Reading	Other
L-V-1	$\chi^2(1) = 8.46$, $p < .01$	$\chi^2(2) = 8.29$, $p < .05$	average vs. good: $\chi^2(1) = 3.70$, $p < .05$
L-V-2	$\chi^2(1) = 7.28$, $p < .01$	n.s.	n.s.
L-V-3	$\chi^2(1) = 11.67$, $p < .01$	$\chi^2(2) = 8.84$, $p < .05$	gr.6 average vs. gr.6 good: $\chi^2(1) = 4.36$, $p < .05$
L-V-4	$\chi^2(1) = 11.47$, $p < .001$	n.s.	n.s.
L-V-5	$\chi^2(1) = 24.92$, $p < .0001$	n.s.	n.s.
L-V-6	n.s.	$\chi^2(2) = 12.94$, $p < .01$	average vs. good: $\chi^2(1) = 3.68$, $p < .055$
L-V-7	n.s.	n.s.	n.s.
L-V-8	$\chi^2(1) = 14.92$, $p < .0001$	n.s.	gr.3 good vs. gr.3 poor: $\chi^2(1) = 3.77$, $p < .05$
L-V-20	$\chi^2(1) = 16.33$, $p < .0001$	n.s.	n.s.
L-V-21	$\chi^2(1) = 21.99$, $p < .0001$	$\chi^2(2) = 13.85$, $p < .001$	poor vs. average $\chi^2(1) = 7.05$, $p < .01$ gr.3 poor vs. gr.3 average $\chi^2(1) = 6.75$, $p < .01$

Table N-8. Language Verbalization Items: Sophisticated Responses

	Grade		Reading Ability		Poor–Average		Average–Good	
	χ^2	p	χ^2	p	χ^2	p	χ^2	p
words	8.46	.004	8.29	.02		n.s.	3.70	.05
wd. where	7.28	.007		n.s.		n.s.		n.s.
sentences	11.67	.0006	8.48	.02		n.s.		n.s.
sent. where	11.47	.0007		n.s.		n.s.		n.s.
"tdet"	24.92	.0001		n.s.		n.s.		n.s.
"meff"		n.s.	12.94	.002		n.s.	3.68	.05
not sure	14.92	.0001		n.s.		n.s.		n.s.
John		n.s.	9.06	.01	3.80	.05		n.s.
Jane	7.45	.006		n.s.		n.s.		n.s.
wented	6.89	.009	6.86	.04		n.s.		n.s.
Mary	47.28	.0001	12.54	.002	4.50	.04		n.s.
TV program	24.57	.0001	19.34	.0001		n.s.	5.15	.02
different	16.33	.0001		n.s.		n.s.		n.s.
homonyms	21.99	.0001	13.85	.001	7.05	.008		n.s.
Meta (1, 2 & 3 vs. 4)	34.45	.0001	18.62	.0001	3.80	.05	4.21	.04

Table N-8, cont'd. Language Verbalization Items: Sophisticated Responses

	3:poor–average		3:average–good		3:good–poor		6:poor–average	6:average–good	
sentences		n.s.		n.s.		n.s.	n.s.	4.36	.04
not sure		n.s.		n.s.	3.77	.05	n.s.		n.s.
TV program		n.s.	6.80	.009		n.s.	n.s.		n.s.
breakfast		n.s.		n.s.		n.s.	n.s.	3.75	.05
different		n.s.		n.s.	4.25	.04	n.s.		n.s.
homonyms	6.75	.009		n.s.		n.s.	n.s.		n.s.
Meta (1, 2 & 3 vs. 4)		n.s.	4.94	.03		n.s.	n.s.		n.s.

Table N-9. Language Verbalizations Items: Correlation with Reading Ability

Item	Grade 3				Grade 6			
	r	p	partial	p	r	p	partial	p
words	.14	n.s.	.09	n.s.	.29	.007	.24	.02
wd. where	.14	n.s.	.11	n.s.	.26	.01	.20	.05
sentences	.17	n.s.	.11	n.s.	.35	.001	.23	.03
sent. where	.20	.05	.21	.04	.10	n.s.	.07	n.s.
"tdet"	.31	.004	.27	.01	.03	n.s.	−.03	n.s.
"meff"	.27	.01	.20	.05	.40	.001	.20	.005
"stone"	.33	.002	.36	.001	.09	n.s.	.06	n.s.
not sure	.37	.001	.34	.004	.05	n.s.	.01	n.s.
best word	.22	.04	.15	n.s.	.32	.003	.29	.007
John	.23	.03	.26	.014	.11	n.s.	.12	n.s.
Jane	.09	n.s.	.05	n.s.	.05	n.s.	.14	n.s.
wented	.27	.01	.26	.02	.09	n.s.	.11	n.s.
Mary	.20	.05	.11	n.s.	.37	.001	.34	.002
dessert	.21	.04	.15	n.s.	.05	n.s.	.12	n.s.
TV program	.43	.001	.33	.003	.20	.05	.20	.05
toothpaste	.21	.04	.16	n.s.	.18	n.s.	.08	n.s.
by man	.13	n.s.	.07	n.s.	.15	n.s.	.16	n.s.
cash	−.003	n.s.	−.003	n.s.	.15	n.s.	.12	n.s.
breakfast	−.05	n.s.	−.04	n.s.	.20	.04	.15	n.s.
different	.31	.004	.25	.02	.27	.01	.26	.02
homonyms	.37	.001	.30	.006	.22	.03	.10	n.s.

Table N-10. Memory Verbalization Items: Chi-Square Results

Item	Grade	Reading Ability	Other Significant
M-V-1	n.s.	n.s.	n.s.
M-V-2	$\chi^2(1) = 17.24$, $p < .001$	n.s.	n.s.
M-V-3	$\chi^2(1) = 23.29$, $p < .0001$	n.s.	n.s.
M-V-4	$\chi^2(1) = 6.97$, $p < .01$	n.s.	n.s.
M-V-5	n.s.	$\chi^2(2) = 6.44$, $p < .05$	n.s.
M-V-6	$\chi^2(1) = 8.73$, $p < .01$	n.s.	n.s.
M-V-7	$\chi^2(1) = 17.27$, $p < .0001$	$\chi^2(2) = 9.41$, $p < .01$	n.s.
M-V-8	$\chi^2(1) = 8.33$, $p < .01$	n.s.	n.s.

Table N-11. Memory Verbalization Items: Correlations With Reading Ability

	Grade 3		Grade 6	
Measure	r	partial r	r	partial r
M-V-1	−.0081	−.2290	.0103	−.0996
M-V-2	−.1356	−.1628	.2732**	.2806**
M-V-3	.2046*	.0948	.1224	.0871
M-V-4	.1515	.2084*	.1716	.0685
M-V-5	.0919	.0730	.4005***	.2179*
M-V-6	−.0862	−.0835	.3444**	.2113*
M-V-7	.2311*	.2080*	.2701**	.2013*
M-V-8	.0263	−.0689	−.0189	−.0358

*$p < .05$ **$p < .01$ ***$p < .001$

Table N-12. Attention Verbalization Items: Chi-Square Results

Item	Grade	Reading Ability	Other Significant
A-V-1	$\chi^2(1) = 36.62$, $p < .0001$	$\chi^2(2) = 7.26$, $p < .05$	n.s.
A-V-2	n.s.	n.s.	n.s.
A-V-3	$\chi^2(1) = 5.70$, $p < .05$	$\chi^2(2) = 6.89$, $p < .05$	gr. 6 average vs. gr. 6 good: $\chi^2(1) = 6.19$, $p < .01$
A-V-4	$\chi^2(1) = 38.78$, $p < .0001$	n.s.	gr. 3 poor vs. gr. 3 good: $\chi^2(1) = 6.95$, $p < .01$ gr. 3 good vs. gr. 6 poor: $\chi^2(1) = 8.71$, $p < .01$ gr. 6 average vs. gr. 6 good: $\chi^2(1) = 3.75$, $p < .05$
A-V-5	n.s.	$\chi^2(2) = 7.07$, $p < .05$	poor vs. average: $\chi^2(1) = 3.68$, $p < .055$
A-V-6	$\chi^2(1) = 11.05$, $p < .001$	$\chi^2(2) = 11.53$, $p < .01$	average vs. good: $\chi^2(1) = 8.23$, $p < .01$; gr. 3 average vs. gr. 3 good: $\chi^2(1) = 7.20$, $p < .01$
A-V-7	$\chi^2(1) = 5.66$, $p < .05$	n.s.	n.s.
A-V-8	$\chi^2(1) = 4.01$, $p < .05$	n.s.	n.s.

Table N-13. Attention Verbalization Items: Correlations with Reading Ability

	Grade 3		Grade 6	
Measure	*r*	partial *r*	*r*	partial *r*
A-V-1	.2191*	.1877	.3303**	
A-V-2	.0818	.0680	.0841	.1209
A-V-3	.2274*	.1591	.3414**	.3779***
A-V-4	.3130**	.2178*	.1378	.0829
A-V-5	.4063***	.3051**	.2156*	
A-V-6	.1880	.1090	.1685	.1474
A-V-7	.0914	.0910	.1042	.1133
A-V-8	.0996	.1116	.0514	.0178

*$p < .05$ **$p < .01$ ***$p < .001$

Table N-14. Developmental Processes: Computed Scores

	Performance	Verbalization
Language		
Grade	$F(1,138) = 102.58$ $p < .001$	$F(1,138) = 110.49$, $p < .001$
Reading Ability	$F(2,138) = 54.98$, $p < .001$	$F(2,138) = 26.51$, $p < .001$
Grade × Reading Ability	n.s.	n.s.
Memory		
Grade	$F(1,138) = 171.86$, $p < .001$	$F(1,138) = 58.72$, $p < .001$
Reading Ability	$F(2,138) = 17.66$, $p < .001$	$F(2,138) = 5.77$, $p < .005$
Grade × Reading Ability	n.s.	n.s.
Attention		
Grade	$F(1,138) = 233.58$, $p < .001$	$F(1,138) = 91.51$, $p < .001$
Reading Ability	$F(2,138) = 19.65$, $p < .001$	$F(2,138) = 15.77$, $p < .001$
Grade × Reading Ability	$F(2,138) = 4.14$, $p < .05$	n.s.

Table N-15. Developmental Processes: Meta Categorizations

Language	
Overall	$\chi^2(15) = 94.23, p < .0001$
Grade	$\chi^2(1) = 34.45, p < .0001$
Reading Ability	$\chi^2(2) = 18.62, p < .0001$
Poor vs. average	$\chi^2(1) = 3.80, p < .05$
Average vs. good	$\chi^2(1) = 4.21, p < .05$
Gr. 3 average vs. Gr. 3 good	$\chi^2(1) = 4.91, p < .05$
Memory	
Overall	$\chi^2(15) = 83.05, p < .0001$
Grade	$\chi^2(1) = 45.78, p < .0001$
Reading Ability	$\chi^2(2) = 8.01, p < .05$
Attention	
Overall	$\chi^2(15) = 108.52, p < .0001$
Grade	$\chi^2(1) = 67.16, p < .0001$
Gr. 3 good vs. gr. 6 poor	$\chi^2(1) = 12.02, p < .001$

APPENDIX O

Language: Performance Items

L-P-1: Vocabulary.

Each child completed the Vocabulary subtest of the Gates- MacGinitie Reading Test. Standard scores were used, with a mean of 50 and a standard deviation of 10 at each grade.

L-P-2 to L-P-4: Production of Language Concepts.

The items were read out loud, and then the children were told to begin. In addition, the experimenter said: "It does not matter what letter, what word, or what sentence you write. I just want one of each." Begin. Time = 2 minutes.

Scoring: letter = 1 point, word = 1 point, sentence = 1 for capital + 1 for period + 1 for grammatically correct and meaningful sentence.

1. On the line below, print *one letter*. (L-P-2)
2. On the line below, print *one word*. (L-P-3)
3. On the line below, print *one sentence*. (L-P-4)

L-P-5: Recognition of Letter.

The instructions were read aloud, and then the children were allowed to begin. Time = 1 minute. Scoring: One point was given for each letter marked.

L-P-6: Recognition of Word.

The instructions were read aloud, and then the children were allowed to begin. Time = 1 minute. Scoring: One point was given for each word correctly marked.

L-P-7: Recognition of Word.

The instructions were read aloud, and then the children were allowed to begin. Time = 1 minute. (Stimuli from Downing & Oliver, 1974.) Scoring: One point was given for each word marked correctly.

(L-P-5)
Put an "X" on each box that has only *one letter* in it.

M	4	Δ	CAT
DOG	□	P	3
MAT	K	2	Δ
7	□	CAN	H
L	BALL	Δ	8
THE HAT	C	DOLL	10
MAN	4	Q	A LOG

(L-P-6)
Put an "X" on each box that has only *one word* in it.

M	4	Δ	CAT
DOG	□	P	3
MAT	K	2	Δ
7	□	CAN	H
L	BALL	Δ	8
THE HAT	C	DOLL	10
MAN	4	Q	A LOG

(L-P-7)
Put an "X" beside each line that has only *one word* on it.

_____ cat
_____ skip and jump
_____ We go to school every day.
_____ television
_____ He had some ice cream.
_____ dog
_____ mother and father
_____ chesterfield
_____ They went to the zoo.
_____ car
_____ airplane
_____ big bad wolf
_____ automobile
_____ fish 'n chips
_____ fire
_____ The dog ran very fast
_____ hippopotomus
_____ He played with the ball.
_____ sun
_____ hide and seek

L-P-8 & L-P-9: Recognition of Sentence.

The instructions were read aloud, and then the children were allowed to begin. Time = 1 minute. Scoring: One point was given for each sentence marked; number of errors (incomplete, phrases) also were noted.

(L-P-8 and L-P-9)
Put an "X" in front of each group of words that makes *a sentence.*

_____ The white house is on the corner.
_____ Only the boys who go to that school.
_____ The big black dog.
_____ After school the children ran home.
_____ He ran.

_____ Before leaving for school, the children.
_____ At the corner, she stopped.
_____ The old woman wore a black hat.
_____ The horse that won the race.
_____ After the party, the children who were in the play.

L-P-10: Recognition of Word/Nonword.
Each child was asked to make three word/nonword judgments, with one point being given for each correct response.

L-P-11: Recognition and Correction of Grammatical Errors.
Each child was asked if 10 sentences were grammatically correct (some sentences from Dennis, unpublished). One point was given for each correct judgment. In addition, if the sentence was incorrect and was recognized as such, the child was asked if he or she could make it a good sentence. One point was given for each correction. The total possible score was 16.

L-P-12: Active/Passive Transformations.
Each child was asked if a series of pairs of sentences meant the same thing. If each member of the pair was judged to mean different things, the child was asked to make the transformation correctly. One point was given for each correct judgment and an additional point was given for each correction, with a total possible score of six.

L-P-13: Production of Homonyms.
Each child was asked to produce two meanings for five homonyms. One point was given for each correct and different meaning, with a possible total of 10. (In the case where three meanings were possible [e.g., to, too, two] credit was given only for the first two.)

L-P-14: Latency of word/nonword judgment.
The length of time (in seconds log 10) to make the word/nonword judgment to "tdet" was recorded.

L-P-15: Latency of word/nonword judgment.
The length of time (in seconds log 10) to make the word/nonword judgment to "meff" was recorded.

L-P-16: Latency of word/nonword judgment.
The length of time (in seconds log 10) to make the word/nonword judgment to "stone" was recorded.

APPENDIX P

Memory: Performance Items

M-P-1: Recall: Immediate

The instructions and words were read aloud, and then the children were allowed to begin. Time = 2 minutes. Turn page.

The instructions were read aloud. The experimenter also said: This is not a spelling test, so write down whatever you think the word looks like. Time = 2 minutes. Scoring: One point for each correct word.

(M-P-1)

You will have two minutes to study these words. Try to remember as many as you can.

box

captain

pound

dinner

problem

car

dress

evening

oil

chair

prince

door

(M-P-1, cont'd)

Write down as many words as you can remember.

1. ________________
2. ________________
3. ________________
4. ________________
5. ________________
6. ________________
7. ________________
8. ________________
9. ________________
10. ________________
11. ________________
12. ________________

M-P-2: Recall: Delay.

The instructions and words were read aloud, and then the children were allowed to begin. Time = 2 minutes. Turn page. The children were told to think about the words. Time = 1 minute. Turn page. The instructions were read aloud, and then the children were allowed to begin. Scoring: One point for each correct word.

(M-P-2)

Try to remember these words.

book

dream

class

horse

bed

egg

cloud

king

building

land

job

city

(M-P-2, cont'd)

Think about the words.

(M-P-2, cont'd)

Write down as many words as you can remember.

1. ______________________________
2. ______________________________
3. ______________________________
4. ______________________________
5. ______________________________
6. ______________________________
7. ______________________________
8. ______________________________
9. ______________________________
10. ______________________________
11. ______________________________
12. ______________________________

M-P-3: Recall: Interference.

The instructions and words were read aloud, and then the children were allowed to begin. Time = 2 minutes. Turn page. The children were told to do as many math problems as possible. Third-grade children were told to start on the second line. Time = 1 minute. Turn page. The instructions were read aloud, and then the children were allowed to begin. Scoring: One point for each correct word.

(M-P-3)

Try to remember these words.

bridge

camp

dollar

cook

husband

ice

salt

rose

peace

clothes

round

party

(M-P-3, cont'd)

Do as many math problems as you can.

Add

24	72	82	66	56	23
36	46	24	37	14	38
52	57	31	18	38	82

345	723	421	321	254	222
231	111	422	344	244	223

(M-P-3, cont'd)

Write down as many words as you can remember.

1. ______________________
2. ______________________
3. ______________________
4. ______________________
5. ______________________
6. ______________________
7. ______________________
8. ______________________

9. ______________________________

10. ______________________________

11. ______________________________

12. ______________________________

M-P-4: Recall: Clusterable List.

The instructions and words were read aloud, and then the children were allowed to begin. Time = 2 minutes. Turn page. The instructions were read aloud, and then the children were allowed to begin. Time = 2 minutes. Scoring: One point for each correct word.

(M-P-4)

Try to remember these words.

eye

ear

face

bear

dog

bird

daughter

brother

son

brown

red

yellow

(M-P-4, cont'd)

Write down as many words as you can remember.

1. ______________________________

2. ______________________________

3. ______________________________

4. ______________________________

5. ______________________________

6. ____________________
7. ____________________
8. ____________________
9. ____________________
10. ____________________
11. ____________________
12. ____________________

M-P-5: Recall of Telephone Number.

After M-V-2, each child was asked to recall a telephone number that had been mentioned. One point was given for each number in the correct order that was recalled, with a total possible of seven.

M-P-6: Recall of Pictures, Story.

After M-V-3, each child was asked to recall the pictures used in the item. One point was given for each picture named, with a total possible of seven.

M-P-7: Recall of Clusterable Pictures.

After M-V-5, each child was asked to recall the pictures used in that item. One point was given for each picture recalled, with a total possible of seven.

M-P-8: Recall of Names.

After M-V-8, each child was asked to recall the names of the children at the birthday party. One point was given for each name recalled, with a total possible of eight.

APPENDIX Q

Attention: Performance Items

A-P-1: Matching: Letters.

The instructions were read aloud, and then the children were allowed to begin. Time = 1 minute. Scoring: One point was given for each correct match.

A-P-2: Matching: Words.

The instructions were read aloud, and then the children were allowed to begin. Time = 1 minute. Scoring: One point was given for each correct match.

A-P-3: Matching: Sentences.

The instructions were read aloud, and then the children were allowed to begin. Time = 2 minutes. Scoring: One point was given for each correct match.

A-P-4: Use of Cues of Clustering (Conceptual).

The instructions were read aloud, and then the children were allowed to begin. Time = 3 minutes. Scoring: Four points for each category (one point for each additional word put in a proper group).

(A-P-4)

The following words are all mixed up. Put the words back in their proper groups by putting some in the first house, some in the second house, and some in the last house.

hat, apple, pear, red, coat, yellow, shoes, blue, peach, green, gloves, cherry

A-P-5: Use of Cues for Clustering (Rhyming).

The words were read aloud, and then the children were allowed to begin. Time = 3 minutes. Scoring: Four points for each category (one point for each additional word put in a proper group).

(A-P-1)

In each row, put an "X" on the *two* boxes that are the *same.*

F	Ⅎ	F	ꟻ
d	b	b	p
y	g	p	g
A	ᗉ	A	ᗆ
O	C	G	C
ə	e	o	e
H	L	H	I

(A-P-2)

In each row, put an "X" on the *two* boxes that are the *same*.

CAP	CAT	HAT	CAT
DOT	DOG	DOG	FOG
MAT	MAP	HAT	MAT
TAN	CAN	CAT	CAN
TALL	BALL	BALL	BAT
DOLL	TOLL	DOLL	DOT
MAN	MAT	CAN	MAN

(A-P-3)

In each box, put an "X" in front of the *two* sentences that are the *same.*

_____ The white house is on the corner. _____ The white horse is on the corner. _____ The white house is on the corner. _____ The white horse ran to the corner.
_____ After school, the children played ball. _____ After school, the children ran home. _____ At school the children played ball. _____ After school, the children ran home.
_____ The girl walked by the store. _____ The girl went into the store. _____ The girl walked into the store. _____ The girl went into the store.
_____ The dog ran after the cat. _____ The dog ran down the street. _____ The cat ran after the dog. _____ The dog ran after the cat.
_____ The horse ran through the field. _____ The horse ran around the field. _____ The horse ran through the field. _____ The house was by the field.

(A-P-5)

hat, fan, tap, pan, cat, lap, can, cap, man, rat, map, pat

A-P-6 & A-P-7: Important Units.

The experimenter said: Listen very carefully as I read the next question out loud. [Read to the end of #1.] Remember that each sentence starts with a capital and ends with a period. Make sure you underline the entire sentence if you think it is important.

Then the children were allowed to begin. Time = 3 minutes. Scoring: 1 point for each important unit. (You have decided that this Saturday would be a good day for the hiking trip since the weather is supposed to be good. You would like to leave at o'clock in the morning unless it is raining. You want to meet your friend on the corner of Main and Elm Streets and leave from there.) Number of other sentences.

(A-P-6 and A-P-7)

Suppose that you and your friend have been planning a hiking trip for a long time. You both like hiking very much, but you haven't been able to do much lately. Last summer you did a lot of hiking together in the woods near your home. Now you would like to go back there to see the changes of the season. You have decided that this Saturday would be a good day for the hiking trip since the weather is supposed to be good. You would like to leave at 9 o'clock in the morning unless it is raining. The long-range weather forecast promises good weather. You have been listening to the weather man on the radio that you got for your last birthday. You want to meet your friend on the corner of Main and Elm Streets and leave from there. This corner is about halfway between your home and your friend's home. It should be a good place to meet.

A-P-8: Production of Note.

#2 was read aloud, and then the children were allowed to begin. Time = 3 minutes. Scoring: One point for each of hiking (not camping) trip, Saturday, 9 a.m., corner of Main and Elm, unless raining (good weather).

(A-P-8)

Suppose that you must leave your friend a note giving him all the important information.

1) Underline each sentence that gives you important information that you must tell your friend. Do not underline any sentence that gives you information that your friend does not need to know for the hiking trip.

2) Suppose you have only a small piece of paper and you are in a hurry so you cannot write a long note. In the space below, write the shortest

note possible, with all the important information that your friend will need to know.

A-P-9 to A-P-12: Search Tasks and Search Patterns.

The child was presented with two search tasks, one redundant (A-P-9, target letters in squares) and one non-redundant (A- P-11). The child was told to find and mark as many target symbols as possible in the time limit (90 seconds). The number correct was noted as a performance measure. (Total = 18 for each task.) The children also were observed as they did the search tasks. They were given two points for a systematic search, one point for a search pattern that varied from systematic to unsystematic or vice versa, and no points for an unsystematic search. The search tasks follow.

(A-P-9 and A-P-10)
Redundant

(A-P-11 and A-P-12)
Nonredundant

				A		G		M				
Q	W	E	R	T	Y	U	I	O	P	L	K	J
H	G	F	D	S	A	Z	X	C	V	B	N	M
Z	A	Q	W	S	X	C	D	E	R	F	V	B
G	T	Y	H	N	M	J	U	I	K	L	O	P
M	N	B	V	C	X	Z	A	S	D	F	G	H
J	K	L	P	O	I	U	Y	T	R	E	W	Q
L	P	M	K	O	N	J	I	B	H	U	V	G
Y	C	F	T	X	D	R	Z	S	E	A	W	Q
P	O	I	U	Y	T	R	E	W	Q	A	S	D
F	G	H	J	K	L	Z	X	C	V	B	N	M
Z	A	Q	W	S	X	C	D	E	R	F	V	B
G	T	Y	G	N	M	J	U	I	K	L	O	P

APPENDIX R

Language: Verbalization Items

Question L-V-1: In school we learn all about words and sentences. I bet you know an awful lot about words and sentences. What can you tell me about words?

Scoring:

0: unable to say anything about words, use them a lot, easy and hard words

1: physical characteristics such as containing letters, length, presence of vowel, read them, spell them, different types, meanings, tell people things.

2: physical characteristics plus meaning

Question L-V-2: Where do we use words?

Scoring:

0: don't know, everywhere, at school, in work, etc.

1: in writing such as sentences, books, etc., in talking, to say things

2: in communicating to others in both "talking" and "writing"

Question L-V-3: What can you tell me about sentences?

Scoring:

0: unable to say anything or irrelevant comment, something that you do in school

1: physical characteristics such as composed of words, capital and period, verb and nouns, an example of a complete sentence or just "complete," meaning, complete thought, tells you something, makes sense

2: physical characteristics plus meaning

Question L-V-4: Where do we use sentences?

Scoring:
0: don't know, in school, in work
1: talking
2: any written form such as books, stories, speeches, reading

Question L-V-5: Once there was a little boy named Johnny who was much younger than you. He didn't know nearly as much as you do about words and sentences. When his teacher asked him to write either a word or a sentence, he often got his paper marked wrong. Suppose that one day his teacher asked you to help him with his words and sentences. You asked him to write a word on his paper. Johnny wrote "tdet" and showed it to you. Would you mark it right or wrong? (The latency of the response was measured in seconds.) What makes you think it is/is not a word?

Scoring:
0: yes, no reason, sounds like a word, looks right, think it means something, or has other characteristics of words such as letters
1: no, no reason, never seen it before, never heard it before, no such word, doesn't look right, doesn't sound right, spelled wrong
2: no, "td" can't go together, doesn't mean anything, doesn't make sense.

Question L-V-6: "Meff." would you mark it right or wrong? What makes you think it is/is not a word?

Scoring:
0: yes, no reason, sounds like a word, looks right, think it means something, heard someone use it, has letters
1: no, no reason or not sure, never seen it before, no such word, doesn't look right, spelled wrong, mixed up, doesn't sound right
2: no, doesn't mean anything or make sense, couldn't use it in a sentence.

Question L-V-7: "Stone." Would you mark it right or wrong? What makes you think it is/is not a word?

Scoring:
0: no, no reason, spelled wrong
1: yes, no reason, just is or irrelevant reason, spelled right, looks right, seen before
2: yes, relation to real object (stones are rocks, can throw them), use word, hear it a lot, such a word, meaning, see it in books, learn how to spell it, learned it in relation to object, not just a bunch of letters put together, label for something

Question L-V-8: Suppose that you were not sure whether to mark one of Johnny's words right or wrong. Is there anything you could do so that you would be sure?

Scoring:

0: no, guess, just check
1: ask somebody, think back, ask Johnny more about it, use it, sound out, check dictionary
2: ask and check dictionary

Question L-V-9: Which one is Johnny's best word? Why? What makes it the best word?

Scoring:

0: other than stone
1: stone, no reason, just is, longer, attribute meaning to all words
2: stone, can use it, more common, meaning

The following 10 questions were presented under the same cover story that was given for the previous questions. The items were designed to assess the children's knowledge of the concept of sentence and grammatical acceptability. Therefore, a question was asked after each sentence to determine whether or not the child was aware of the problem in the sentence and if so, if he or she could correct it.

Question L-V-10 to L-V-19: Then you asked Johnny to write one sentence. He wrote:

(10) John park to went.
(11) Jane played with her friends.
(12) After school, Bill wented home.
(13) Before Mary could enter the contest.
(14) My favorite dessert is radios with cream.
(15) My favorite TV program are Gunsmoke.
(16) My favorite toothpaste is Crest.
(17) I paid the money by the man.
(18) I gave the cash to the girl.
(19) My favorite breakfast is eggs with bacon.

[After each sentence, the child was asked the following questions.] Would you mark it right or wrong? [A performance measure.] What makes it a sentence/ not a sentence?

Scoring for 11, 16, 18, and 19:

0: no, no reason or change something that's right, refuses to guess or their change does not affect the acceptability of the sentence
1: yes, no reason, but something must be added, looks OK, sounds OK, not wrong, long, spelled right

2: yes, all the parts there, all go together, proper sentence, right order, repeats sentence or paraphrases, capital and period, means something, makes sense, complete thought

Scoring for all other sentences (10, 12, 13, 14, 15, 17):

0: yes, sounds right, makes sense, no idea, refuses to guess or identifies the mistake as a "mistake," gives characteristics of sentence

1: no, but identifies wrong element, not a sentence, not long enough or don't know, sounds wrong, doesn't make sense

2: no, explains actual mistake

Note: For wrong sentences, each child was asked the following: If you were Johnny, how would you change it to make it a good sentence? [A performance measure.]

The following item was designed to assess the children's knowledge of the flexibility of language. An attempt was made to have the children recognize and make changes to the structure of language before asking them about ways to make their language sound different.

Question L-V-20: Once there were two friends, Ernie and Bernie. Bernie likes to say things that mean the same thing as whatever Ernie says, but Bernie always tries to make it sound different. For example, if Ernie says, "Bill hit the ball," Bernie would say, "The ball was hit by Bill." Both Ernie and Bernie said the same thing, but it sounded different. A) Suppose that Ernie said, "The cake was eaten by Jack," and Bernie said, "Jack ate the cake." (B) Suppose that Ernie said, "John hit Mary," and Bernie said, "Mary hit him back." C) Suppose that Ernie said, "Joe carried the box," and Bernie said, "The box was carried by Joe." D) Suppose that Ernie said, "The dog chased the cat," and Bernie said, "The cat chased the dog." After each pair, the child was asked the following questions: Did Bernie say the same thing as Ernie? Did they mean the same thing? For B and D, if there was a negative response, the child was asked: What should Bernie have said? After all four pairs, the child was asked: Can you always say the same thing in different ways? How do you make the same idea sound different?

Scoring:

0: no idea

1: switch the words around, use different words that express the same meaning, add more words (for more explanation)

2: any combination of two or more ways

Question L-V-21: Sometimes it is best to explain what you mean by a word by using it in a sentence. You've done that before, haven't you? Sometimes, though, you can give two different meanings for a word. For example, if I said "blue"

and asked you to make up a sentence to explain the meaning of the word, you could say, "the sky is blue" or "The wind blew through the trees" or "He is blue today" (for blue mood or sad). All of those answers would be right. Now I want you to give me a sentence that would explain the meaning of: to (too, two), here (hear), ate (eight), hour (our), there (their, they're). Now give me another sentence that would make ____________ mean something different. (If a second meaning was not produced by the child, a second sentence was given by the experimenter.) After all pairs, the child was asked: How come you can give me two meanings for each of the words we have talked about so far? If I kept giving you words, do you think that you could always give me two different meanings for every word? Why do you think you could/could not?

Scoring:

0: yes, just could, no reason why not, smart
1: no, no reason or wrong reason, not smart, too hard, no but usually can, some words hard, don't know them, might make mistakes, words might not make sense
2: no, some words don't have a sound-alike twin, some words have only one meaning

APPENDIX S

Memory: Verbalization Items

Question M-V-1: Suppose you wanted to phone your friend and someone told you the phone number. Like, suppose I said the number was 555-8643. Would it make any difference if you called right away after you heard the number or if you got a drink of water first? Why?

Scoring:
- 0: wouldn't matter, no reason, drink first, thristy
- 1: wouldn't matter because of use of strategy such as write it down or memorize first or could remember
- 2: call first, no reason or irrelevant reason, call first because might forget

Question M-V-2: What do you do when you want to remember a telephone number?

Scoring:
- 0: remember it, nothing, ask someone to remember, ask again
- 1: write it down, say it over and over, repetition, memorize, keep in head
- 2: combination of write and repetition, cluster numbers

Note: At this time each child was asked the following: Do you remember the number that I told you? What was it? [For each number remembered in proper order (1); a performance measure]

The following item was designed to assess the child's knowledge of the demands of the data on the ability to recall. In this particular case, the knowledge that was being assessed was whether or not the child was aware that hearing a story about pictures would provide a structure that would make the pictures easier to remember.

Question M-V-3: The other day I showed these pictures to other boys and girls your age. I asked one girl to learn them so that she could tell me what they were later when she couldn't see them any more. And I showed the same pictures to another girl, but also told her a story about the pictures. [Experimenter puts down each picture as its depicted object is mentioned.]

A man gets up out of *bed* and gets dressed, putting on his best *shirt* and *shoes*. Then he sits down at the *table* for breakfast. After breakfast he takes his *dog* for a walk. Then he puts on his *hat* and walks out the *door* to go to work.

I told the girl who heard this story that she was supposed to learn the pictures so she could tell me what they were later when she couldn't see the pictures. She didn't have to tell me the story, just the pictures. Do you think the story made it easier or harder for the girl to remember the pictures? Why?

Scoring:
- 0: harder, any reason
- 1: easier, no reason
- 2: easier, use of theme, pictures in story, in order

Note: At this point, each child was asked the following: Can you tell me the names of the pictures? What were they? [For each picture (1); performance measure.]

Question M-V-4: [The stimuli are nine $3 \times 4\frac{1}{2}$-inch colored pictures randomly arranged in a 3×3 matrix. They are potentially clusterable into three conceptual categories: animals = squirrel, chick, frog; food = grapes, carrots, corn; clothing = mittens, coat, shoes.]

Now suppose I wanted you to learn these pictures. You could do anything you wanted with the pictures. You might want to move them around, for example. You would have three minutes to look and study, but then I would take the pictures away and ask you what pictures you learned. What would you do to learn these pictures?

Scoring:
- 0: nothing
- 1: look at them a lot, read or sound out words, put in head, use of cues, repetition, story, partial clustering, study, self-test
- 2: clustering

Question M-V-5: Why would you do it that way? [Referring to above item.] Is there anything else that you could do?

Scoring:
- 0: no

1: use of second strategy, partial clustering, repetition, etc.
2: use of clustering as a second strategy

Note: At this point, each child was asked the following: Do you learn everything this way? [Pictures removed.] Can you tell me the names of the pictures? What were they? [For each picture (1); a performance measure.]

Question M-V-6: Suppose you lost your jacket while you were at school. How would you go about finding it?

Scoring:
0: just look around, no systematic search
1: use of relevant cues but not systematic, look in yard, ask people to help
2: systematic search, go back along path till you find it, remember where you were, what you were doing

Question M-V-7: Anything else that you could do? [Referring to the above item.]

Scoring: For each possible way, 1 point: i.e., look alone, ask people such as principal or teacher, ask kids to help look, lost and found or announcements, go back to remember:
O: 0–1 way
1: 2–3 ways
2: 4–5 ways

Question M-V-8: One day two friends went to a birthday party and they met eight children whom they didn't know before. I'll tell you the names of the children they met: Bill, Fred, Jane, Sally, Anthony, Jim, Lois, and Cindy. After the party one friend went home and the other went to practice a play that he was going to be in. At the play practice he met seven other children he didn't know before, and their names were Sally, Anita, David, Marie, Jim, Dan and Fred. At dinner that night, both children's parents asked them the names of the children they met at the birthday party that day. Which friend do you think remembered the most, the one who went home after the party or the one who went to practice in the play, where he met some more children? Why?

Scoring:
0: play practice
1: party then home, no reason or irrelevant reason
2: party, then home, less names, no interference

Note: At this point, each child was asked the following: How many of the names of the children at the birthday party do you remember? Can you name them? [For each correct name (1); a performance measure.]

APPENDIX T

Attention: Verbalization Items

Question A-V-1: Suppose you had a lot of books to choose from, like in a library. How would you decide which one you wanted to read?

Scoring:

0: just look or pick one, have someone else choose, one that you have read before, etc.

1: use of possible cues such as title, table of contents, pictures, compare books by simple cues such as length, title, pictures, read whole thing

2: compare books by skimming, read parts, read summary, advanced strategy plus search strategy such as use of card catalogue, find type of book, book strategies for several books to choose the best one

Question A-V-2: Suppose that you were a teacher and you wanted to tell the parents of all the kids in your class whether or not their child was a good reader. What would be the best way to decide who was a good reader?

Scoring:

0: don't know, spell words

1: read out loud or one other reason such as reading hard books (do not count irrelevant reasons such as good speller), reading test

2: combination of aloud and test

Question A-V-3: Why would that be the best way?

Scoring:

0: don't know, just would, only way, irrelevant reason

1: read aloud so that you can hear them, listen for expression, interested in hard things, test to see if they understand
2: combination of reasons, aloud and test

Question A-V-4: Suppose that you wanted to put words in alphabetical order. How would you do it?

Scoring:
0: don't know, the alphabet would tell you
1: first letter, second letter
2: beyond second letter

Question A-V-5: Suppose I gave you a list of words and asked you to put all the rhyming words together in their proper groups. How would you do it?

Scoring:
0: don't know, middle or first sounds
1: look at them, think about them, look at last letters, match or say the words
2: read words to see if rhyme, sound at end (listen to end)

Question A-V-6 & Question A-V-7: Redundancy task: The child was presented with two search tasks, one redundant (target letters all in squares) and one nonredundant (no difference between target and nontarget letters). The child was told to find and mark as many target symbols as possible in the time limit (90 seconds). The number correct was noted as a performance measure. The child then was asked the following questions. How did you find so many so quickly?

Scoring:
0: don't know, irrelevant reason, knew letters, good eyes
1: just looked, looked fast all around, in circles
2: looked in systematic way, down each row, for redundant only, use of squares

Question A-V-8: Which one was easier? [Referring to the above tasks] What made it easier?

Scoring:
0: second, any reason or no difference
1: first, use of irrelevant cues, length, or don't know
2: first, use of shapes or interference of first on second, or confusing letters (G and C), use of squares

APPENDIX U

Computed Scores: Correlation Matrices

Correlation matrixes of computed scores for Grades 3 and 6 appear on the following pages.

Grade 3

	RC	AGE	L-P	A-P	M-P	D-P	S-P	C-P	L-V	A-V	M-V	D-V	S-V	C-V	IQ
RC	1.00	−0.02	0.64	0.57	0.45	0.74	0.71	0.61	0.58	0.44	0.15	0.23	0.30	0.52	0.40
AGE		1.00	−0.13	0.00	−0.00	−0.04	−0.08	−0.05	−0.09	−0.06	−0.07	0.10	−0.13	0.42	-0.33
L-P			1.00	0.60	0.40	0.70	0.70	0.65	0.55	0.42	0.13	0.24	0.38	0.60	0.41
A-P				1.00	0.48	0.55	0.53	0.46	0.49	0.38	0.06	0.28	0.29	0.39	0.42
M-P					1.00	0.46	0.41	0.37	0.18	0.36	0.24	0.17	0.26	0.22	0.36
D-P						1.00	0.76	0.72	0.50	0.34	0.14	0.12	0.17	0.54	0.35
S-P							1.00	0.97	0.58	0.41	0.24	0.25	0.35	0.79	0.51
C-P								1.00	0.51	0.34	0.22	0.21	0.30	0.83	0.48
L-V									1.00	0.51	0.31	0.47	0.54	0.40	0.91
A-V										1.00	0.22	0.24	0.46	0.34	0.36
M-V											1.00	0.36	0.36	0.16	0.34
D-V												1.00	0.38	0.21	0.08
S-V													1.00	0.33	0.42
C-V														1.00	0.41
IQ															1.00

(cont'd)

Grade 6

	RC	AGE	L-P	A-P	M-P	D-P	S-P	C-P	L-V	A-V	M-V	D-V	S-V	C-V	IQ
RC	1.00	0.06	0.73	0.52	0.51	0.75	0.79	0.67	0.48	0.41	0.42	0.20	0.44	0.57	0.57
AGE		1.00	0.18	−0.17	−0.13	0.03	−0.07	-0.08	0.18	−0.19	0.00	−0.05	0.04	−0.02	−0.28
L-P			1.00	0.47	0.44	0.69	0.64	0.57	0.61	0.31	0.37	0.06	0.36	0.62	0.45
A-P				1.00	0.46	0.35	0.42	0.34	0.40	0.47	0.19	0.09	0.28	0.37	0.49
M-P					1.00	0.34	0.50	0.43	0.25	0.31	0.33	0.29	0.47	0.54	0.56
D-P						1.00	0.28	0.60	0.38	0.40	0.24	0.12	0.35	0.50	0.38
S-P							1.00	0.96	0.43	0.40	0.43	0.18	0.44	0.76	0.44
C-P								1.00	0.37	0.34	0.37	0.21	0.36	0.80	0.34
L-V									1.00	0.30	0.45	0.16	0.27	0.37	0.28
A-V										1.00	0.22	0.13	0.36	0.10	0.24
M-V											1.00	0.16	0.51	0.35	0.40
D-V												1.00	0.11	0.19	0.31
S-V													1.00	0.29	0.41
C-V														1.00	0.31
IQ															1.00

In all cases, $df = 70$.

For $p = .05$, $r = 0.23$.

For $p = .01$, $r = 0.30$.

Note: RC refers to a standardized test of reading comprehension and IQ refers to a nonverbal measure of intelligence.

Author Index

Subject Index

Springer Series in Language and Communication

Continued from page ii